Robert Schulze

From hot lattice QCD to cold quark stars

AF004684

Robert Schulze

From hot lattice QCD to cold quark stars

A quasiparticle approach

Südwestdeutscher Verlag für Hochschulschriften

Impressum/Imprint (nur für Deutschland/only for Germany)
Bibliografische Information der Deutschen Nationalbibliothek: Die Deutsche Nationalbibliothek verzeichnet diese Publikation in der Deutschen Nationalbibliografie; detaillierte bibliografische Daten sind im Internet über http://dnb.d-nb.de abrufbar.

Alle in diesem Buch genannten Marken und Produktnamen unterliegen warenzeichen-, marken- oder patentrechtlichem Schutz bzw. sind Warenzeichen oder eingetragene Warenzeichen der jeweiligen Inhaber. Die Wiedergabe von Marken, Produktnamen, Gebrauchsnamen, Handelsnamen, Warenbezeichnungen u.s.w. in diesem Werk berechtigt auch ohne besondere Kennzeichnung nicht zu der Annahme, dass solche Namen im Sinne der Warenzeichen- und Markenschutzgesetzgebung als frei zu betrachten wären und daher von jedermann benutzt werden dürften.

Verlag: Südwestdeutscher Verlag für Hochschulschriften GmbH & Co. KG
Heinrich-Böcking-Str. 6-8, 66121 Saarbrücken, Deutschland
Telefon +49 681 37 20 271-1, Telefax +49 681 37 20 271-0
Email: info@svh-verlag.de

Approved by: Dresden, TU, Diss., 2011

Herstellung in Deutschland:
Schaltungsdienst Lange o.H.G., Berlin
Books on Demand GmbH, Norderstedt
Reha GmbH, Saarbrücken
Amazon Distribution GmbH, Leipzig
ISBN: 978-3-8381-3128-3

Imprint (only for USA, GB)
Bibliographic information published by the Deutsche Nationalbibliothek: The Deutsche Nationalbibliothek lists this publication in the Deutsche Nationalbibliografie; detailed bibliographic data are available in the Internet at http://dnb.d-nb.de.

Any brand names and product names mentioned in this book are subject to trademark, brand or patent protection and are trademarks or registered trademarks of their respective holders. The use of brand names, product names, common names, trade names, product descriptions etc. even without a particular marking in this works is in no way to be construed to mean that such names may be regarded as unrestricted in respect of trademark and brand protection legislation and could thus be used by anyone.

Publisher: Südwestdeutscher Verlag für Hochschulschriften GmbH & Co. KG
Heinrich-Böcking-Str. 6-8, 66121 Saarbrücken, Germany
Phone +49 681 37 20 271-1, Fax +49 681 37 20 271-0
Email: info@svh-verlag.de

Printed in the U.S.A.
Printed in the U.K. by (see last page)
ISBN: 978-3-8381-3128-3

Copyright © 2012 by the author and Südwestdeutscher Verlag für Hochschulschriften GmbH & Co. KG and licensors
All rights reserved. Saarbrücken 2012

Kurzfassung

Ein thermodynamisches Modell des Quark-Gluon-Plasmas mit Quasiteilchenfreiheitsgraden basierend auf Hard-Thermal-Loop-Selbstenergien wird vorgestellt. Es stellt ein Bindeglied zwischen einem bereits etablierten, phänomenologischen Quasiteilchenmodell [Pes00, BKS07a], welches aus ersterem durch einer Reihe von Näherungen folgt, sowie QCD, aus welcher ersteres mittels des Cornwall-Jackiw-Tomboulis-Formalismus und einer speziellen Parametrisierung der laufenden Kopplung abgeleitet werden kann, dar.

Beide Modelle ermöglichen die Extrapolation von Monte-Carlo-Gitterberechnungen der QCD-Zustandsvariablen bei kleinen chemischen Potentialen zu großen Nettobaryonendichten mit bemerkenswert ähnlichen Ergebnissen und werden so genutzt, um Zustandsgleichungen für Schwerionenkollisionsexperimente am SPS und bei FAIR ebenso wie für Quark- und das Innere von Neutronensternen anzugeben. Eine Mischphasenkonstruktion erlaubt die Verbindung der SPS-Zustandsgleichung zum Hadronenresonanzgasmodell.

Eine Erweiterung auf den schwach wechselwirkenden Sektor wird vorgelegt und allgemeine Argumente bezüglich der Stabilität und Bindung von kompakten stellaren Objekten abgeleitet. Die Ergebnisse von Extrapolationen aktuell verfügbarer Gitterrechnungen [Baz09, Bor10b] legen die Existenz von reinen Quarksternen nicht nahe. Allerdings könnte Quarkmaterie in einer Mischphase in Kernen von Neutronensternen existieren.

Abstract

A thermodynamic model of the quark-gluon plasma using quasiparticle degrees of freedom based on the hard thermal loop self-energies is introduced. It provides a connection between an established phenomenological quasiparticle model [Pes00, BKS07a] – following from the former using a series of approximations – and QCD – from which the former is derived using the Cornwall-Jackiw-Tomboulis formalism and a special parametrization of the running coupling.

Both models allow for an extrapolation of first-principle QCD results available at small chemical potentials using Monte-Carlo methods on the lattice to large net baryon densities with remarkably similar results. They are used to construct equations of state for heavy-ion collider experiments at SPS and FAIR as well as quark and neutron star interiors. A mixed-phase construction allows for a connection of the SPS equation of state to the hadron resonance gas.

An extension to the weak sector is presented as well as general stability and binding arguments for compact stellar objects are developed. From the extrapolation of the most recent lattice results [Baz09, Bor10b] the existence of bound pure quark stars is not suggested. However, quark matter might exist in a hybrid phase in cores of neutron stars.

for Constanze, Rahel and Elias

Contents

1 Introduction **9**
    1.1 Standard model of particle physics 9
    1.2 Quantum Chromodynamics 10
    1.3 The quark-gluon plasma 13
    1.4 Heavy-ion collisions and quasiparticles 15
    1.5 Outline of the work 16

2 Derivation of the hard thermal loop quasiparticle model **17**
    2.1 Self-consistent approximations of many-body systems 17
    2.2 Evaluation of the traces 19
    2.3 Application to QCD 20
    2.4 Properties of the HTL self-energies; Landau damping 23
    2.5 Investigation of the dispersion relations 25
    2.6 The HTL grand canonical potential 27
    2.7 Effective coupling 28
    2.8 Entropy density 29
    2.9 The pressure 32
    2.10 Solution of the flow equation 35
    2.11 Integration of the mean field pressure 38
    2.12 Interaction measure – connection to lattice QCD 39

3 Analytic investigation of the model **41**
    3.1 Asymptotic dispersion relations and quark restmasses 41
    3.2 Net quark density 43
    3.3 Partial pressures 45
    3.4 Mean field pressure contribution 47
    3.5 Entropy and energy density 48

4 The effective quasiparticle model **51**
    4.1 Necessary approximations 51
    4.2 Comparison with lattice results 53
    4.3 Extrapolation to nonzero chemical potential 55
    4.4 Check of model consistency 56
    4.5 Expansion for small chemical potentials 60
    4.6 Results at small temperatures 63
    4.7 The explicit dependence of the asymptotic masses on the chemical potential .. 63
    4.8 Wrap-up 65

5 Equation of state for heavy-ion collider experiments **67**
    5.1 Comparison with lattice results 67
    5.2 Contributions of collective modes 68
    5.3 Pressure susceptibilities 70
    5.4 Extrapolation to nonzero chemical potential 71
    5.5 Equation of state for SPS 76
    5.6 Equation of state for FAIR 80
    5.7 Application in QCD sum rule calculations 80
    5.8 Wrap-up 81

6 Compact stellar objects **83**
    6.1 Gravitation and general relativity 84
    6.2 The Tolman-Oppenheimer-Volkoff equations 85
    6.3 Including the weak sector 85
    6.4 Extrapolation to nonzero chemical potential 87
    6.5 State variables at zero temperature 92
    6.6 Equation of state 95
    6.7 Analytic investigation of the TOV equations 97
    6.8 Pure quark stars 100
    6.9 Hybrid approaches 102
    6.10 Wrap-up 109

7 Summary **111**

A Evaluation of Matsubara sums **113**

B Mathematical relations **117**
    B.1 Imaginary part of the logarithm 117
    B.2 Derivative of Arg and arctan 118

C List of derivatives ... **119**
- C.1 Derivatives of the HTL thermal masses 119
- C.2 Derivatives of the eQPM asymptotic masses 120

D Coefficients of the flow equations ... **123**
- D.1 Effective quasiparticle models 123
- D.2 HTL quasiparticle model 124

Bibliography ... **133**

1 Introduction

1.1 Standard model of particle physics

Within the standard model of particle physics, matter consist of two classes of fundamental particles: leptons and quarks (and their corresponding antiparticles), see Fig. 1.1. These spin-$1/2$ particles (fermions) interact via three fundamental forces: weak, electromagnetic and strong interaction – in order of rising strength. In addition, gravity is present as fourth fundamental interaction.

With the exception of gravity, the interactions can be described by *gauge field theories*. Within these theories an interaction is described through the exchange of gauge bosons (particles with integer spin, here spin-1). The quantum field theoretical treatment of gravity has not yet reached a satisfying level to be included into the standard model – but quite a few proposals (superstrings, M-theory, loop quantum gravity, etc.) exist.

The strength of an interaction is given by its *coupling constant*, which – in spite of its name – is not constant but rather depends on the energy-momentum scale (or, alternatively, the distance). This is due to the quantum vacuum – also in spite of its name – not being empty, but instead consisting of quantum fluctuations. For instance, in the case of Quantum Electrodynamics (QED), virtual electron-positron pairs acting as virtual dipoles appear, screening charged particles and thus reducing the interaction strength. The screening is more effective in reactions on small energy-momentum scales (larger distances), thus the QED coupling constant increases with the increasing energy and momentum.

At very high energies ($\sim 10^{15}$ GeV), the coupling strength of weak, strong and electromagnetic interaction seem to merge. This gives rise to the hope that all fundamental forces can be unified into one which was split during the first moments after the Big Bang. With the exception of the electroweak interaction as unification of electromagnetic and weak forces, this seems not yet to be achieved in a convincing manner.

The focus of this thesis is the description of properties of strongly interacting matter. The corresponding gauge field theory is called Quantum Chromodynamics (QCD).

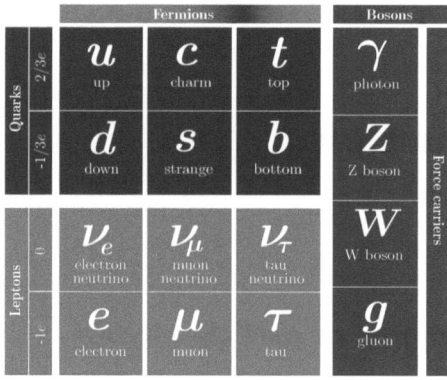

Figure 1.1: Elementary particles according to the standard model (Higgs and antiparticles omitted). Created after [PS95].

1.2 Quantum Chromodynamics

In analogy to QED, which very accurately describes electromagnetic interactions as exchange of photons between electrically charged particles, strong interactions arise from an exchange of gluons between quarks that carry an additional charge property. While the electromagnetic force couples to particles with a property electric charge with just one single aspect (positive or negative elementary charge e), the strong interacting couples to particles with a property with three aspects, which are analogized by the three elementary colors red, green and blue, and the property named color charge [Gre64, Nam74] (hence the prefix *chromo*).

This analogy has been very successful, because it illustrates the QCD phenomenon "confinement", i.e. color charged particles cannot be isolated singularly, by the requirement that quarks form "white" (neutral color charge) compounds. Employing the notion of constituent[1] quarks, this can be achieved either by a quark-antiquark pair with opposite colors (called *meson*, e.g. composed of a red and an anti-red quark) or three quarks of a different color each (called *baryon*, e.g. composed of one red, one green and one blue quark). The matter surrounding us is essentially made up of protons, neutrons and electrons. While the electron is one of the leptons and therefore fundamental, the proton and the neutron are baryons composed of *up* and *down* constituent quarks (Fig. 1.2).

While QED has a simple Abelian gauge group structure, namely $U(1)$, leading to only one gauge boson (the photon), the non-Abelian $SU(3)$ gauge group of QCD contains eight gluons, each carrying a unique color-anticolor combination (except white). In this way, the gluons are color-charged quanta themselves and consequently interact not only among each other by

---

[1] The constituent quarks used to understand hadron spectra have to be distinguished from the current quarks of the gauge theory. Although they carry the same quantum numbers such as isospin, strangeness, etc., their masses may differ.

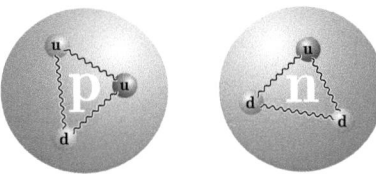

Figure 1.2: Naive pictorial representation of a proton (p) and a neutron (n) as composites of up (u) and down (d) quarks. Wiggly lines schematically indicate gluons as carriers of the strong force. Derived from [KSS06].

exchanging further gluons – which in turn interact via even more gluons – but also with the virtual quarks and gluons of the vacuum.

As a consequence, the strong coupling strength g is small for short quark interdistances (or high momentum transfers) and increases with growing distance (or decreasing momentum transfer leading to gluon "inflation" and antiscreening of the quarks). This effect is called *asymptotic freedom*.[2]

Even though a mathematical proof is still required, it is assumed that quarks and gluons can never be observed as isolated particles. The energy spent for the separation of two color-charged particles leads to another pair being created in between.

QCD as a theory is based on the classical Lagrangian (cf. [PDG06], p. 110 and p. 319)

$$\mathcal{L}_{\text{QCD}} = \sum_q \bar{\psi}_q^i (i\gamma^\mu (D_\mu)_{ij} - \delta_{ij} m_q) \psi_q^j - \frac{1}{4} F_{\mu\nu}^a F_a^{\mu\nu} + \mathcal{L}_{\text{gauge}} + \mathcal{L}_{\text{FP}}, \qquad (1.1)$$

where $\mu, \nu = 0 \ldots 3$ are Lorentz indices, $i, j = 1 \ldots 4$ are Dirac spinor indices, and $a = 1 \ldots 8$ is the adjoint color index of the gluon color states. The sum over all quark flavors $q \in \{u, d, c, s, t, b\}$ is given explicitly; for the remaining indices the Einstein sum convention has to be followed.

The quark fields ψ_q (color triplets) refer to current quarks. They are coupled minimally to the gauge sector by the covariant derivative

$$(D_\mu)_{ij} = \delta_{ij} \partial_\mu + ig \frac{(\lambda_a)_{ij}}{2} A_\mu^a, \qquad (1.2)$$

where the A_μ^a represent the gauge fields (gluons) and the λ_a are the generators of the local $SU(3)$ gauge group in the fundamental representation.[3]

---

[2]For the discovery of this phenomenon in 1973 – which led to the widespread acceptance of the non-Abelian QCD in the following years – D. Gross, D. Politzer and F. Wilczek have received the Nobel prize in 2004. Asymptotic freedom has been verified, e.g. in deep inelastic scattering experiments. It is due to this effect that, rather than coupling strength, the notion *running coupling* is more accurately being employed for g.

[3]The standard representation of λ_a are the Gell-Mann matrices. They are the three-dimensional extension of the Pauli matrices. They form the Lie algebra of $SU(3)$ with commutation relations $[\lambda_a, \lambda_b] = 2i f_{abc} \lambda^c$, where f_{abc} are the fully antisymmetric structure constants.

The pure Yang-Mills term $\mathcal{L}_{\text{YM}} := -\frac{1}{4} F^a_{\mu\nu} F_a^{\mu\nu}$ describes the gluons just as any other gauge boson. For the gluons the field strength tensor is given by

$$F_a^{\mu\nu} = \underbrace{\partial^\mu A_a^\nu - \partial^\nu A_a^\mu}_{\text{analogous to QED}} + \underbrace{g f_{abc} A^{b\mu} A^{c\nu}}_{\text{gluon-gluon interaction}} \quad (1.3)$$

As a result of the additional terms, $\mathcal{L}_{\text{YM}}$ contains expressions trilinear and quadrilinear in the gluon fields A_a^μ leading to three- and four-gluon interactions.

The contribution $\mathcal{L}_{\text{gauge}}$ fixes the still remaining gauge degree of freedom and $\mathcal{L}_{\text{FP}}$ takes care of possibly occurring unphysical degrees of freedom by introducing Fadeev-Popov ghost fields. A very insightful, deeper look at those phenomena is given by [GTP11].

The current quark masses m_q and the coupling strength g have to be adjusted to physical observables. Due to renormalization within the quantized theory, the quantities m_q and g entering the classical Lagrangian (1.1) become subject to a redefinition resulting in a scale dependence. The same holds for matter and gauge sector fields.

By performing an expansion in terms of its small coupling constant ($\alpha_{em} \approx 1/137$ at asymptotically small energy scales) the equations for QED scattering processes can be solved up to a certain energy scale where it diverges [PS95]. Conform with asymptotic freedom, the situation is reversed for the QCD running coupling g which is smaller than unity only at very high energy, density or momentum scales. For larger g, non-perturbative methods are needed in order to solve the QCD equations of motion. One way is to discretize space and time and apply Monte-Carlo sampling methods. This approach is commonly called lattice QCD. Such lattice calculations are, however, still limited to small net baryon densities. This is due to a numerical sign problem, i.e. highly oscillatory integrands appearing with the introduction of nonzero chemical potentials.

For vanishing quark masses the QCD Lagrangian (1.1) is chirally symmetric, meaning that both left and right handed world are fully decoupled. Lattice results are often obtained using unphysically large quark masses and therefore need to be extrapolated towards the *chiral limit*, i.e. to the physical mass scales. This is in addition to the extrapolation towards the thermodynamic continuum accounting for errors introduced due to the discretization of space and time.

Even for small values of g, naive perturbation theory for a strongly interacting system can fail at nonzero temperature T despite the use of modern expansion methods [CH98]. This is due to the energy scale introduced by the temperature which leads to expansion terms $\sim gT/p$ [BP90a, BP90b]. These are no longer of order g, but can be of order unity for a typical momentum scale p of particles in a heat bath of temperature T. The usual relation of the order of the loop expansion and powers of g is lost: effects of leading order in g arise from every order in the loop expansion and one cannot arrange different contributions according to powers of g anymore. Consequently, in order to still be able to apply perturbative methods, all these terms need to be taken into account which can be done by resummation techniques.

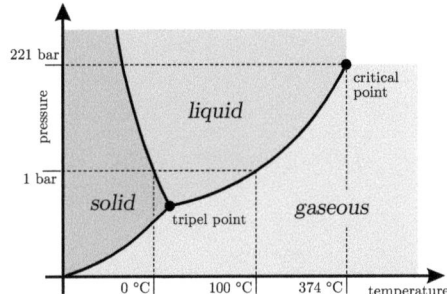

Figure 1.3: A part of the phase diagram of water, showing its three major phases and the phase transitions. (*source*: www public domain)

1.3 The quark-gluon plasma

We often experience matter in three different phases: solid, fluid and gaseous. However, certain materials have a much richer phase structure. For instance, water can assume twelve or more different ice phases [LFK98], the usual fluid form or become vapor. These phases transform into each other with the change of exterior conditions such as pressure and temperature as depicted in Fig. 1.3. These phase transitions are often combined with tremendous changes in material properties like compressibility, transparency or electrical conductivity.

If water is brought to sufficiently high temperatures it turns into yet another state: a plasma consisting of ions and quasifree electrons. Since the transition happens slowly through ionization of single molecules by collisions it is not considered a phase change. Still a plasma shows new collective effects such as screening and plasma oscillations and is therefore often considered as fourth state of matter.

The plasma phase can also be reached through compression, by which electrons are released from their orbitals and form a degenerate quantum plasma. This phase transition is observed e.g. when a star collapses into a White Dwarf (which is a stable compact stellar object due to the sufficiently large degeneracy pressure of these electrons).

The nuclei of water molecules consist of protons and neutrons which are again formed by constituent quark triplets (Fig. 1.2) interacting – as outlined in the previous sections – through an exchange of gluons. Quarks and gluons are confined within the nucleons. Just as, after heating and compressing it sufficiently, the constituents of water molecules form a plasma, this strongly interacting matter is presumed to deconfine, i.e. transit into a phase with much less correlation often assuming quarks and gluons as freely roaming. This phase is called the *quark-gluon plasma* (QGP). Note that although the question about the nature of the transition (first/second order phase transition or just a crossover) actually depends on the number of active quark flavors N_f as well as their masses, the use of the terminology "transition" is widely accepted.

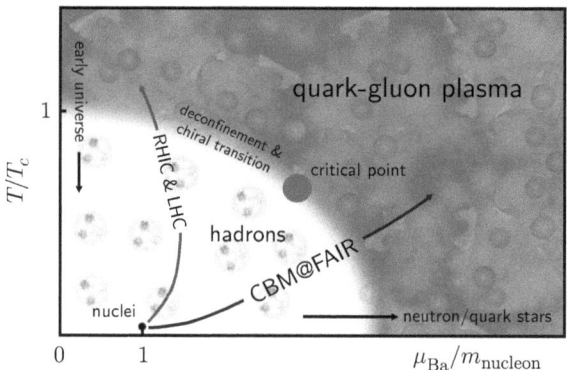

Figure 1.4: Schematic plot of the phase diagram of strongly interacting matter for constant volume V and variable net baryon density or baryo-chemical potential μ_{Ba} scaled by a typical nucleon mass and temperature T scaled by the transition temperature at $\mu_{Ba} = 0$. The white region for low μ_{Ba} and T represents the hadronic regime, the red region the quark-gluon plasma with the deconfinement and chiral transition in between. The arrows depict areas accessible to current and future experiments and the evolutionary path of the universe. Neutron star matter occupies the region of low temperature and large baryon density. For the means of abbreviations see text. (derived from GSI publicity plot)

Fig. 1.4 shows the presumed phase diagram of strongly interacting matter. The red region is occupied by deconfined matter, where quarks and gluons are believed to represent the relevant degrees of freedom. The white region represents hadronic matter. At the common boundary, strongly interacting matter undergoes the deconfinement transition.

Within this thesis we assume that the restoration of chiral symmetry (i.e. chiral condensate $\langle \bar{\psi}\psi \rangle \to \approx 0$) occurs simultaneously or sufficiently close to the deconfinement phase transition. Consequently, a possible quarkyonic phase speculated to exist between deconfinement and chiral phase transition [MP07] is neglected.

For small chemical potentials and $N_f = 2+0$, $2+1$ the deconfinement transition is a simple crossover. At the critical point it becomes a real phase transition of second order turning into first order for larger values of μ_{Ba}. The transition temperature of the crossover at vanishing chemical potential is commonly dubbed pseudocritical or simply critical temperature T_c. The common starting point of the experiments marks hadrons or nuclei in our environment. Not shown is a possible triple point with the color superconducting phase predicted by some authors [RW00].

The arrows in Fig. 1.4 indicate the path of the early universe, the area relevant for neutron and possible quark stars and the scope of some present and future experiments outlined within the following chapter.

1.4 Heavy-ion collisions and quasiparticles

In order to investigate the properties of deconfined matter enormous energy scales are necessary. While they currently cannot be provided continuously, it is possible to accelerate nuclei to relativistic velocities, let them collide and register all relevant exit particles in a detector. By comparison with simulations the data (angle, momentum, time, ...) of these particles can be used to reconstruct the processes which occurred during and after the collision, e.g. if deconfined matter has been produced.

Collision experiments, aimed at producing the quark-gluon plasma, have been performed at the Alternating Gradient Synchrotron (AGS) and Relativistic Heavy Ion Collider (RHIC) at Brookhaven National Laboratories as well as the the Super Proton Synchrotron (SPS) at CERN in Geneva. Two new experiments are indicated in Fig. 1.4. LHC is the Large Hadron Collider now in use at CERN in Geneva; GSI SIS300 is related to FAIR in Darmstadt. The latter is aimed to investigate the region of high net baryon densities at sufficiently high energies.

If after the collision of the nuclei, the quarks and gluons of the emerging fireball become deconfined, they form a QGP which is assumed to behave as a fluid or plasma or gas with features to be specified. In order to predict the dynamics of the fireball during this phase and assuming a thermally equilibrated system, hydrodynamics can be employed. This amounts to solving the equations of energy-momentum and current conservation [LL06]

$$T^{\mu\nu}{}_{;\mu} = 0 \qquad \text{and} \qquad (nu^\mu)_{;\mu} = 0 \qquad (1.4)$$

with the energy-momentum tensor of an ideal fluid $T^{\mu\nu} = (e+p)u^\mu u^\nu - pg^{\mu\nu}$, where e is the energy density, p stands for the pressure and n denotes the net density of particles of the respective matter emerging from hadrons at the deconfinement transition while u^μ is the 4-velocity. Here, $g^{\mu\nu} = \text{diag}(1,-1,-1,-1)$ is the metric tensor of Minkowski space.

Since the problem is under-determined, it is necessary – as one important input for the hydrodynamic description – to provide an interrelation of the state quantities e, p and n, for instance of the form $e = e(p,n)$, i.e. an *equation of state* (EOS) for strongly interacting matter, in particular the quark-gluon plasma. As the thermodynamic potential contains all of the information on the bulk properties of matter in local equilibrium, the goal is to derive an expression for, e.g., the grand canonical potential Ω from the QCD Lagrangian and the associated Feynman rules (i.e. propagators, self-energies, etc.). The EOS then follows as a direct consequence.

Given the problems in solving the QCD equations of motion by perturbation theory and lattice calculations, especially in the region of high baryon densities as mentioned in Section 1.2, a different ansatz is needed. The interpretation of the QGP constituents as noninteracting quasiparticles using effective masses is one possible approach to a solution of the problem.

A quasiparticle is an elementary excitation of a system. If chosen in a sensible way, other excitations of the system can be described by the presence of multiple quasiparticles. In certain limits the interaction between multiple quasiparticle species can turn out to be negligible, giving the possibility to investigate properties of the many-body system by examining properties of the individual, non-interacting species.

The concept of quasiparticles is one of the few established techniques in simplifying a quantum mechanical many-body problem. It has proven to be very successful within the field of condensed matter physics but its scope of application is not limited to it. Recent work [And10a, And10b, Str10] underline the suitability of the concept for the QGP. This thesis makes use of the quasiparticle picture in order to describe the thermodynamics of the quark-gluon plasma.

As for most many-body systems, the QGP is expected to possess two classes of quasiparticles: the first one corresponding to actual particles, the properties of which are modified by the interactions within the plasma; the other class representing quanta of collective excitation modes in the system. For the QGP, the longitudinal gluon (or plasmon) mode and the plasmino [Kli81, Wel82] constitute such collective excitations [YHM95].

1.5 Outline of the work

In Chapter 2 the hard thermal loop (HTL) quasiparticle model is derived from a 2-loop effective action and a special parametrization of the running coupling. The properties of the HTL self-energies, propagators and dispersion relations are investigated and the issues of thermodynamic consistency and crossing characteristics are addressed in detail.

In Chapter 3, the properties of the thermodynamic quantities are studied and the asymptotic dispersion relations are introduced, allowing for a connection to a previously used quasiparticle model and – in the reverse direction – provide a coherent path for the improvement of the latter. This connection is finalized in Chapter 4. In addition, the results of the established model for the most recent lattice results are given and discussed. A modification neglecting the explicit μ-dependence of the thermal masses is studied as well.

The differences between the established and the HTL model as improvement of the former are discussed in Chapter 5. The HTL model is then used to construct an equation of state for heavy-ion collider experiments at high net baryon densities.

With the goal to obtain an equation of state for compact stellar objects with a high-density quark phase, the weakly interacting sector is included into the quasiparticle model in Chapter 6. General arguments concerning existence and properties of pure quark stars as well as neutron stars with quark cores are presented. Actual predictions using the most recent lattice results are provided and discussed in terms of these arguments.

Chapter 7 summarizes the arguments and results of this thesis.

2 Derivation of the hard thermal loop quasiparticle model

2.1 Self-consistent approximations of many-body systems

In order to determine the thermodynamic potential of a relativistic many-body system from its Lagrangian and the associated propagators and self-energies, i.e. its excitation spectrum, in a thermodynamically consistent way, it is necessary to construct the sum of all two particle irreducible (2PI) skeleton diagrams (i.e. diagrams without external lines which do not become disconnected upon cutting of two propagator lines). Employing either the Cornwall-Jackiw-Tomboulis formalism or the Φ-functional approach these graphs are translated into the sum over all self-energy contributions via a functional derivative. The resulting full propagators are then used to determine the thermodynamic potential.

The Φ-functional approach is based on a proposal by Luttinger and Ward [LW60] to derive the thermodynamic potential of non-relativistic, fermionic systems from Feynman graphs, i.e. using propagator expressions. As such this formalism is a translation of a stationarity theorem by Lee and Yang [LY60b] – which expresses the thermodynamic potential Ω in terms of mean occupation numbers – into propagator language.[1] The formalism has also been extended to bosonic [GW65, FW71] and relativistic systems [NC75, VB98].

Only a few years after Luttinger and Ward, it was Jona-Lasinio [Jon64], who emphasized the importance of the *effective action* for discussions of spontaneous symmetry breaking in relativistic particle theory. Consequently, Jona-Lasinio, Dahmen and Tarski [DJ67, DJ69, DJT72] presented a variational formulation of relativistic quantum field theory based on combinatorial analysis. It was reformulated using functional methods and presented with some example applications in a review by Cornwall, Jackiw and Tomboulis [CJT74] which was widely referenced, leading to the term *CJT formalism*.[2] Essentially, the CJT formalism is the result of a generalization of the Φ-functional approach. This relation is elaborated in [Sch07].

---

[1] In fact, it can even be reformulated to use any subset of n-point amplitudes with $n \leq 4$ [dDM64, NC75, Kle82, Car04, Ber04].

[2] The formalism was developed independently within the Soviet science community by Vasil'ev and Kazanskii [VK72, VK73a, VK73b].

The starting point of the CJT formalism is the *effective action*[3] Γ [Ris03]

$$\Gamma[D, S] = I - \frac{1}{2} \left\{ \text{Tr} \left[\ln D^{-1} \right] + \text{Tr} \left[D_0^{-1} D - 1 \right] \right\}$$
$$+ \left\{ \text{Tr} \left[\ln S^{-1} \right] + \text{Tr} \left[S_0^{-1} S - 1 \right] \right\} + \Gamma_2[D, S], \qquad (2.1)$$

where I is the classical action of the system and D and S are the dressed bosonic and fermionic propagators with their respective tree-level equivalents D_0 and S_0. Γ_2 represents the sum over all 2PI diagrams without external lines analogous to the Φ-functional.

The effective action is subject to stationarity conditions

$$\frac{\delta \Gamma}{\delta D} = \frac{\delta \Gamma}{\delta S} = 0 \qquad (2.2)$$

which yield

$$0 = -D^{-1} + D_0^{-1} - 2 \frac{\delta \Gamma_2}{\delta D} \quad \text{and} \quad 0 = S^{-1} - S_0^{-1} - \frac{\delta \Gamma_2}{\delta S} \qquad (2.3)$$

and thus lead to the gap equations

$$\Pi = -2 \frac{\delta \Gamma_2}{\delta D} \quad \text{and} \quad \Sigma = \frac{\delta \Gamma_2}{\delta S}, \qquad (2.4)$$

where gluon and quark self-energies are defined by Dyson's equations

$$\Pi[D] := D^{-1} - D_0^{-1} \quad \text{and} \quad \Sigma[S] := S^{-1} - S_0^{-1}. \qquad (2.5)$$

For translationally invariant systems, where propagators fulfill $\Delta_{(0)}(x, y) = \Delta_{(0)}(x - y)$, an overall factor of the four-dimensional space T/V can be extracted from the trace integrals in (2.1) and it suffices to consider the *effective potential*

$$V_{\text{eff}}[D, S] = -\frac{T}{V} \Gamma[D, S] \qquad (2.6)$$

instead of the effective action [CJT74, Ris03, Bec05, Roe05]. At the stationary point V_{eff} is connected to the grand canonical potential via (see [Bro92], p. 104, or [Riv88])

$$\frac{\Omega}{V} = V_{\text{eff}}. \qquad (2.7)$$

The stationarity condition (2.2) is conveyed from the the effective action so that the thermodynamic potential has to be stationary with respect to a variation of the propagators as well

$$\frac{\delta \Omega}{\delta D} = \frac{\delta \Omega}{\delta S} = 0. \qquad (2.8)$$

---

[3] For more on the concept of the effective action in the framework of symmetry breaking, the interested reader is referred to [PS95] and [Riv88].

The classical action I can also be replaced by a classical potential $U = -IT/V$, which describes the broken symmetries of the system. Since no broken symmetries are considered in this thesis, U is equal to zero and one obtains the expression for the thermodynamic potential

$$\Omega[D, S] = T \left\{ \frac{1}{2}\text{Tr}\left[\ln D^{-1} - \Pi D\right] - \text{Tr}\left[\ln S^{-1} - \Sigma S\right] \right\} + T\Gamma_2[D, S]. \tag{2.9}$$

Since Γ_2 is an infinite sum, these gap equations can presently only be solved approximately by selecting a subset of the skeleton diagrams in Γ_2. The truncated self-energies are then found from (2.4) in an elegant way: The functional derivative with respect to the propagators is graphically represented by simply cutting one propagator line in the skeleton diagrams, keeping in mind all the symmetry factors. From the truncated self-energies, the corresponding truncated dressed propagators follow self-consistently from Dyson's equations (2.5).

Even though truncation introduces approximations, they were shown to obey the fundamental physical laws such as number, energy and momentum conservation [BK61, Bay62]. Therefore, this approach is often called conservation law preserving or symmetry conserving self-consistent approximation scheme. Baym [Bay62] also introduced the notion Φ-*derivable approximations* in reference to the Φ/Γ_2-functional.

2.2 Evaluation of the traces

In the expressions above, Tr contains a trace tr over the discrete indices as well as the integration over the four-dimensional phase space. In order to take effects of finite temperature into account, the energy integration has to be performed using the imaginary time formalism (see [LeB96, Kap89, YHM95] for an introduction). This is done by carrying out a sum over the discrete Matsubara frequencies

$$i\omega_n = \begin{cases} 2ni\pi T & \text{for bosons,} \\ (2n+1)\,i\pi T + \mu & \text{for fermions,} \end{cases} \tag{2.10}$$

where μ denotes the one independent chemical potential[4] of the system. Since the expression for the thermodynamic potential depends on the square of the three-momentum only, i.e. it is rotationally invariant, the momentum integral $(2\pi)^{-3} \int d^3k$ reduces to $(2\pi^2)^{-1} \int dk\, k^2$. Translation invariance gives an overall factor of the three-volume V of the system yielding

$$\text{Tr} \longrightarrow \frac{V}{2\pi^2} \text{tr} \sum_n \int dk\, k^2. \tag{2.11}$$

---

[4]Considering the case of $N_f = 2$ dynamical quark flavors and assuming zero net electric charge as well as equal u and d quark masses, the isospin chemical potential $\mu_I = (\mu_u - \mu_d)/2$ vanishes. Therefore, there is only one independent chemical potential $\mu = \mu_q = \mu_u = \mu_d = \mu_{\text{Ba}}/3$, where μ_{Ba} is the baryo-chemical potential (cf. Fig. 1.4).

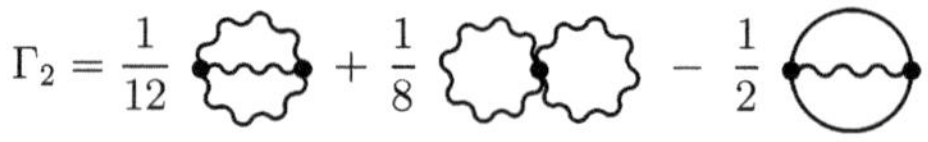

$$\Gamma_2 = \frac{1}{12} \underset{}{\bigotimes} + \frac{1}{8} \underset{}{\bigotimes\!\bigotimes} - \frac{1}{2} \underset{}{\bigoplus} \qquad (2.14)$$

Figure 2.1: Contributions to Γ_2 at 2-loop order; wiggly lines are gluons, solid lines represent quarks.

The Matsubara sum is performed using standard contour integration techniques with details relegated to Appendix A. Applying the result

$$T \sum_{n=-\infty}^{+\infty} f(p_0 = i\omega_n) = -\int_{-\infty}^{+\infty} \frac{d\omega}{\pi} n_B(\omega) \, \text{Im}(f(\omega + i\varepsilon)), \qquad (2.12)$$

where $n_B = (\exp(\beta\omega) - 1)^{-1}$ with $\beta = 1/T$ is the Bose-Einstein statistical distribution function, and an analogous expression for fermions, with opposite sign and the Fermi-Dirac distribution $n_F = (\exp(\beta(\omega - \mu)) + 1)^{-1}$, yields the following expression for the thermodynamic potential from Eq. (2.9):

$$\begin{aligned}\frac{\Omega}{V} &= \text{tr} \int \frac{d^4k}{(2\pi)^4} n_B(\omega) \, \text{Im}\!\left(\ln D^{-1} - \Pi D\right) \\ &+ 2\,\text{tr} \int \frac{d^4k}{(2\pi)^4} n_F(\omega) \, \text{Im}\!\left(\ln S^{-1} - \Sigma S\right) - \frac{T}{V}\Gamma_2,\end{aligned} \qquad (2.13)$$

where the propagators D and S now represent the retarded bosonic and fermionic propagators, respectively. Thus, in the following only retarded propagators and corresponding self-energies are used.

2.3 Application to QCD

In order to apply the CJT formalism to QCD we need to select a subset of graphs in Γ_2, i.e. to truncate Γ_2 at a reasonable perturbative order. Since Γ_2 contains two-particle irreducible diagrams only, there are no 1-loop contributions and the first non-trivial contributions are encountered at 2-loop order. As it turns out, these contributions already contain a rich structure and approximate many features of QCD closely. We include all 2-loop contributions in our choice of Γ_2, which is shown in Fig. 2.1.

Following [Bay62] a self-consistent approximate solution which conserves particle number, energy and momentum can be obtained. This is achieved by first computing both quark and gluon self-energies using the gap equations, i.e. performing a functional variation of Γ_2 with respect to the propagators. This can be interpreted as cutting one propagator line within the Feynman graphs. Taking the prefactors and symmetries into account, the 2-loop contributions to Γ_2 lead to the 1-loop self-energies shown in Fig. 2.2.

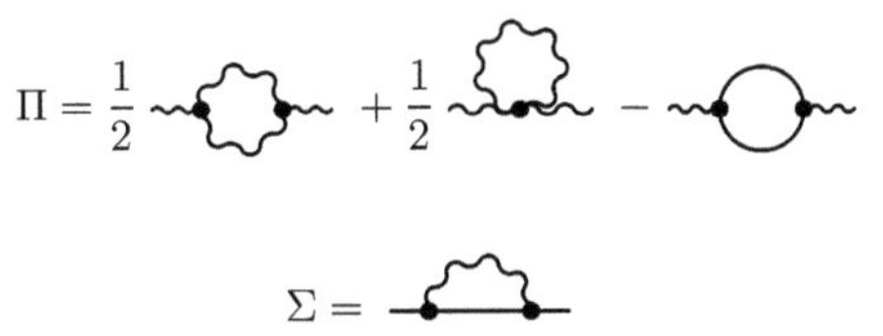

2.3 Application to QCD

$$\Pi = \frac{1}{2}\ \text{\scriptsize(diagram)}\ + \frac{1}{2}\ \text{\scriptsize(diagram)}\ -\ \text{\scriptsize(diagram)} \qquad (2.15)$$

$$\Sigma = \ \text{\scriptsize(diagram)} \qquad (2.16)$$

Figure 2.2: The 1-loop QCD self-energies derived from Γ_2 at 2-loop order.

Although, as a consequence of the truncation, gauge invariance is lost[5], it can be restored by assuming soft external momenta or equivalently Hard Thermal Loops (HTL) in the propagator and self-energy expressions. For 1-loop QCD in the chiral limit[6] the HTL approximation provides gauge invariant self-energies [BP90b]. We follow the conventions of [BIR01][7] and use the HTL self-energies and propagators for gluons and massless quarks given therein:

$$\begin{aligned}
\Pi_{\mu\nu} &= \Pi_\mathrm{T}(\omega,k)\left(\Lambda_\mathrm{T}(\vec{k})\right)_{\mu\nu} - \Pi_\mathrm{L}(\omega,k)\left(\Lambda_\mathrm{L}(\vec{k})\right)_{\mu\nu}, \\
\gamma_0 \Sigma &= \Sigma_+(\omega,k)\ \Lambda_+(\vec{k}) \quad - \quad \Sigma_-(\omega,k)\ \Lambda_-(\vec{k})
\end{aligned} \qquad (2.17)$$

with projectors $\Lambda_T = \delta_{ij} - k_i k_j / k^2$ and $\Lambda_L = k_i k_j \omega^2 / k^4$. Both modes are transverse to k^μ, the denomination transverse and longitudinal is defined in reference to $\vec{k}$. In contrast to the vacuum case, the longitudinal gluon mode which, at zero temperature, is a static mode producing the familiar Coulomb interaction, propagates for nonzero temperature and has to be taken into account [BP90a].

The scalar self-energies are given as

$$\begin{aligned}
\Pi_\mathrm{T}(\omega,k) &= \frac{m_D^2}{2}\left(1 + \frac{\omega^2 - k^2}{k^2}\Pi_\mathrm{L}(\omega,k)\right), \\
\Pi_\mathrm{L}(\omega,k) &= m_D^2\left(1 - \frac{\omega}{2k}\ln\frac{\omega+k}{\omega-k}\right), \\
\Sigma_\pm(\omega,k) &= \frac{\hat{M}^2}{k}\left(1 - \frac{\omega \mp k}{2k}\ln\frac{\omega+k}{\omega-k}\right),
\end{aligned} \qquad (2.18)$$

where $\hat{M} = \hat{M}(T, \mu, g^2)$ is the thermal fermion mass or plasma frequency and $m_D = m_D(T, \mu, g^2)$

---

[5]The gauge dependence of Φ-derivable approximations has been studied in [Arr02]. They were shown to possess a controlled gauge dependence, i.e. the gauge dependent terms emerge at orders higher than the truncation order.

[6]For nonzero quark masses, the quark self-energy is no longer gauge invariant (cf. [Sei07]).

[7]In particular, we also use Coulomb gauge. As shown in [BIR01], ghost contributions can be neglected if transversality of the polarization tensor is manually enforced as in Eqs. (2.17) with 4-dimensionally transverse projectors. For a very detailed account cf. [Sei07].

is the Debye mass. They read

$$
\begin{aligned}
m_D^2 &= \underbrace{\left(\frac{C_b}{3} T^2 + \frac{N_c}{6\pi^2} \sum_{q=1}^{N_f} \mu_q^2 \right)}_{2\tilde{C}_b} g^2, \\
\hat{M}^2 &= \underbrace{\frac{C_f}{8} \left(T^2 + \frac{\mu^2}{\pi^2} \right)}_{\tilde{C}_f} g^2,
\end{aligned}
\qquad (2.19)
$$

where $C_f = (N_c^2 - 1)/(2N_c)$ and $C_b = N_c + N_f/2$. N_c is the number of colors (in this thesis fixed at 3) and $N_f = N_l + N_h$ is the number of present quark flavors. For flexibility in comparison with lattice calculations, we assume $N_l = 2$ light quark flavors (up and down quarks) of equal chemical potential μ and, if present, $N_h = 1$ heavy quark flavor (strange quark) of chemical potential zero[8], so that $\sum_{q=1}^{N_f} \mu_q^2 = N_l \mu^2$.

The different thermal fermion masses of the quark flavors are signified by an index q where necessary. For general discussion and implying the dependence on the respective chemical potential, it suffices the consider the general expression $\hat{M}^2$, applicable for the light quark flavors, as $\hat{M}_s^2 = \hat{M}^2|_{\mu=\mu_s=0}$ follows for a heavy strange quark.

Since the additional HTL approximation impairs self-consistency, the term "approximately self-consistent approximation" has been established. It is worth mentioning that, since only undressed vertices are used, the Ward identities are obviously violated. We follow the reasoning in [BIR01] that the vertex corrections can be implemented self-consistently but are, however, negligible at 2-loop order.

Finally Dyson's equations (2.5) are used to self-consistently determine the dressed propagators

$$
\begin{aligned}
D_T^{-1} &= -\omega^2 + k^2 + \Pi_T, \\
D_L^{-1} &= - k^2 - \Pi_L, \\
S_\pm^{-1} &= -\omega \pm (k + \Sigma_\pm).
\end{aligned}
\qquad (2.20)
$$

---

[8]Generally, the exact strange quark chemical potential μ_s has to be included in the description of the considered plasma. However, if the net strangeness given by a certain initial condition is zero and there is no overall change of net strange quark number (e.g. due to strangeness conservation in strong interaction processes) μ_s vanishes. This constellation with is referred to by a flavor number $N_f = 2 + 1$. It is a good approximation, e.g., for heavy-ion collisions, as proton and neutron are both comprised of u and d quarks only. While $s\bar{s}$-pairs may appear, strangeness conservation could only be violated by weak interactions for which strong interaction time-scales are too short.
In Chapter 6 we will consider a case with $\mu_s \neq 0$ as the above argument does not hold for quark matter in compact stellar objects.

2.4 Properties of the HTL self-energies; Landau damping

This section deals with symmetries and other properties of the real and imaginary parts of the retarded self-energies. The real and the imaginary parts of the HTL self-energies (2.18) are[9]

$$\begin{aligned}
\mathrm{Re}\Pi_T(\omega,k) &= \frac{1}{2}m_D^2\left(\frac{\omega^2}{k^2} - \frac{\omega^2-k^2}{k^2}\frac{\omega}{2k}\ln\left|\frac{\omega+k}{\omega-k}\right|\right), \\
\mathrm{Re}\Pi_L(\omega,k) &= m_D^2\left(1 - \frac{\omega}{2k}\ln\left|\frac{\omega+k}{\omega-k}\right|\right), \\
\mathrm{Re}\Sigma_\pm(\omega,k) &= \frac{\hat{M}^2}{k}\left(1 - \frac{\omega\mp k}{2k}\ln\left|\frac{\omega+k}{\omega-k}\right|\right),
\end{aligned} \quad (2.23)$$

$$\begin{aligned}
\mathrm{Im}\Pi_T(\omega,k) &= \frac{1}{2}m_D^2\frac{\omega^2-k^2}{k^2}\frac{\omega}{2k}\pi\Theta\left(k^2-\omega^2\right)\varepsilon(k), \\
\mathrm{Im}\Pi_L(\omega,k) &= m_D^2\frac{\omega}{2k}\pi\Theta\left(k^2-\omega^2\right)\varepsilon(k), \\
\mathrm{Im}\Sigma_\pm(\omega,k) &= \frac{\hat{M}^2}{k}\frac{\omega\mp k}{2k}\pi\Theta\left(k^2-\omega^2\right)\varepsilon(k),
\end{aligned} \quad (2.24)$$

where $\varepsilon(k)$ is the sign function. The gluon self-energies show the symmetries

$$\begin{aligned}
\mathrm{Re}\Pi_i(-\omega) &= \mathrm{Re}\Pi_i(\omega), \\
\mathrm{Im}\Pi_i(-\omega) &= -\mathrm{Im}\Pi_i(\omega),
\end{aligned} \quad (2.25)$$

i.e. the real parts are symmetric and the imaginary parts are antisymmetric with respect to the energy ω. This can explicitly be seen for a momentum of $k = 0.5T$ in Fig. 2.3. Analogously, the quark self-energies fulfill the parity relations

$$\begin{aligned}
\mathrm{Re}\Sigma_+(-\omega) &= \mathrm{Re}\Sigma_-(\omega), \\
\mathrm{Im}\Sigma_+(-\omega) &= -\mathrm{Im}\Sigma_-(\omega)
\end{aligned} \quad (2.26)$$

as shown for $k = 0.5T$ in Fig. 2.4.

The HTL self-energies do not account for quasiparticle widths, as there is no imaginary part at the poles of the propagator. The nonzero imaginary parts of the self-energies below

---

[9] For the imaginary part of the logarithm of $z = (\omega+k)/(\omega-k)$, retardation $\omega + i\varepsilon$

$$z(\omega+i\varepsilon) = \frac{\omega+i\varepsilon+k}{\omega+i\varepsilon-k} = \frac{\omega+k}{\omega-k} - i\varepsilon\frac{2k}{(\omega-k)^2} \quad (2.21)$$

is decisive, as applying the infinitesimally small imaginary part of $z(\omega+i\varepsilon)$ to Eq. (B.3) gives

$$\mathrm{Im}\ln\frac{\omega+k}{\omega-k} = \pi\epsilon(-k)\Theta(-\frac{\omega+k}{\omega-k}) = -\pi\Theta(k^2-\omega^2)\varepsilon(k) \quad (2.22)$$

while it would be zero for non-retarded z.

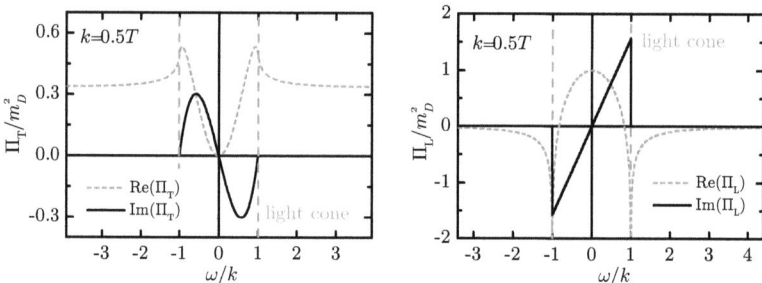

Figure 2.3: The real and imaginary parts of the retarded transverse (left) and longitudinal (right) gluon self-energies scaled by the Debye mass squared m_D^2 (Eq. (2.19)) are shown as functions of the energy ω scaled by the momentum k which is fixed at $k = 0.5T$. The real parts are symmetric with respect to ω, while the imaginary parts are antisymmetric and differ from zero only below the light cone $|\omega| = k$.

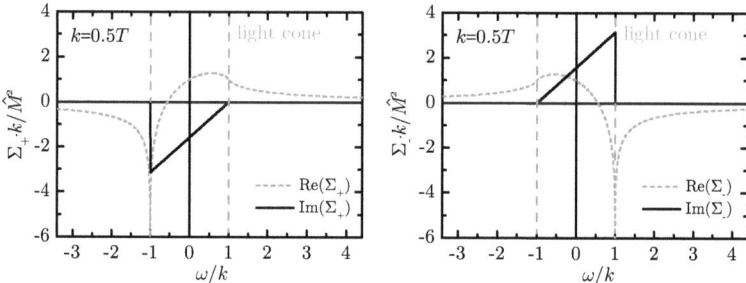

Figure 2.4: The real and imaginary parts of the retarded quark self-energies for the normal (left) and abnormal branch (right) scaled by the plasma frequency squared $\hat{M}^2$ (Eq. (2.19)) are shown as functions of the energy ω scaled by the momentum k which is fixed at $k = 0.5T$. They fulfill the parity relations $\mathrm{Re}\Sigma_+(-\omega) = \mathrm{Re}\Sigma_-(\omega)$ and $\mathrm{Im}\Sigma_+(-\omega) = -\mathrm{Im}\Sigma_-(\omega)$. The imaginary parts of the self-energy are nonzero only below the light cone.

the light cone are due to *Landau damping* (LD). LD is a collective effect caused by energy transfer between the gauge field and plasma particles with velocities close to the phase velocity ("resonant particles"). As this resonance would be spoiled by collisions in a normal fluid, it is a unique feature of collisionless plasmas [ONC99].

Consider particles whose velocity is slightly higher than ω/k prior to an energy transfer. If they gain energy from the gauge field they leave the area of resonance, while, if losing energy to the gauge field, they approach the resonant velocity even closer and can again interact with the gauge field. These particles would effectively transfer energy to the gauge field.

In the opposite case, particles with velocity slightly below ω/k effectively gain energy from the gauge field. Since physical distributions favor states of lower energy, the states of energy loss are usually less populated than the ones which gain energy. Therefore, a net energy transfer

2.5 Investigation of the dispersion relations

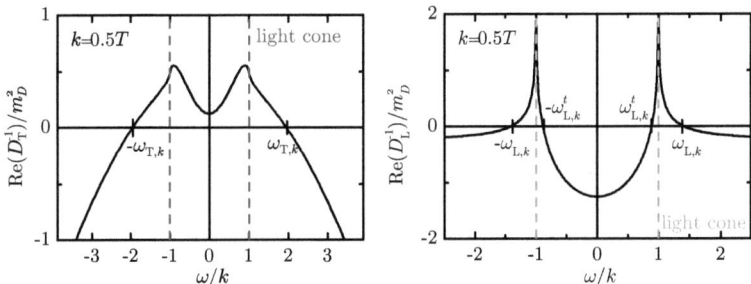

Figure 2.5: The real parts of the inverse gluon propagators $D^{-1}_{\mathrm{T,L}}$ scaled by the Debye mass squared are shown as functions of the energy ω scaled by the momentum k which is fixed at $k = 0.5T$. Both are symmetric with respect to ω. The root of $\mathrm{Re}D^{-1}_{\mathrm{T}}$ determines the dispersion relation $\omega_{\mathrm{T},k}$ for transverse gluons. The root of $\mathrm{Re}D^{-1}_{\mathrm{L}}$ above the light cone indicates the dispersion relation $\omega_{\mathrm{L},k}$ of longitudinal gluons, while the tachyonic dispersion relation $\omega^t_{\mathrm{L},k}$ (below the light cone) is due to Landau damping.

to the particles takes place, damping the gauge field.[10]

Even though the imaginary parts are formally nonzero only below the light cone, retardation leads to an infinitely small contribution even above the light cone, giving a definite sign to the imaginary parts of the self-energies for all ω:

$$\begin{aligned}\varepsilon(\mathrm{Im}\Pi_{\mathrm{T}}(\omega)) &= -\varepsilon(\omega),\\ \varepsilon(\mathrm{Im}\Pi_{\mathrm{L}}(\omega)) &= +\varepsilon(\omega),\\ \varepsilon(\mathrm{Im}\Sigma_{\pm}(\omega)) &\equiv \mp 1.\end{aligned} \quad (2.27)$$

Note that the sign of the imaginary parts above the light cone is solely due to retardation and not related to Landau damping which is found below the light cone only.

2.5 Investigation of the dispersion relations

On-shell (quasi)particles satisfy the dispersion relation $\mathrm{Re}\Delta^{-1} = 0$, where Δ denotes the respective propagator. It is therefore useful to investigate the real part of the inverse retarded HTL propagators as to identify the relevant degrees of freedoms described therein.

Both inverse gluon propagators $D^{-1}_{\mathrm{T,L}}$ are symmetric in the energy domain and have just one positive energy dispersion relation above the light cone: $\omega_{\mathrm{T},k}$ and $\omega_{\mathrm{L},k}$, respectively. This means that – up to the sign – transverse and longitudinal gluons have the same dispersion relations as their anti(quasi)particle counterparts. The additional tachyonic dispersion relation

---
[10]Thus, Landau damping prevents the collisionless plasma from becoming unstable. In contrast, Cherenkov instabilities, i.e. the gauge field gaining energy from the particles, may occur in some non-Maxwellian plasmas where states of higher energy are more populated than states of lower energy, e.g. a beam-plasma system. [TL97]

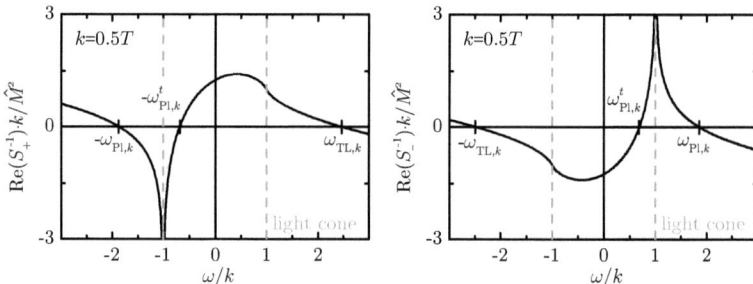

Figure 2.6: The real parts of the inverse quark propagators $S_\pm^{-1}$ scaled by the fermionic mass parameter squared are shown as functions of the energy ω scaled by the momentum k which is fixed at $k = 0.5T$. The real part of neither inverse quark propagator shows any symmetry with respect to ω. However, a parity relation $\mathrm{Re}S_+^{-1}(-\omega) = -\mathrm{Re}S_-^{-1}(\omega)$ holds. The tachyonic dispersion relation $\omega_{\mathrm{Pl},k}^t$ is due to Landau damping.

for longitudinal gluons is related to Landau damping. It is screened by the latter and does not propagate. Fig. 2.5 shows the real parts for fixed momentum $k = 0.5T$.

The inverse quark propagators are not symmetric per se but, as a consequence of Eq. (2.26), satisfy the parity property (cf. Fig. 2.6)

$$\mathrm{Re}S_+^{-1}(-\omega) = -\mathrm{Re}S_-^{-1}(\omega). \tag{2.28}$$

This is a sign of the intricate nature of both propagators: quarks are described by the positive energy dispersion relation of S_+^{-1}, while the dispersion relation of antiquarks is found from the negative energy solution of $\mathrm{Re}S_-^{-1} = 0$. The remaining two dispersion relations represent collective quark excitations: the positive energy dispersion relation of S_-^{-1} describes the plasminos, while the negative energy solution of $\mathrm{Re}S_+^{-1} = 0$ represents antiplasminos. Again, a tachyonic solution appears within the regime of Landau damping.

The evolution of the roots of the real part of the inverse retarded propagators as a function of the momentum k gives the dispersion relations $\omega_{i,k}$. It is one of the difficulties of the subject at hand that these dispersion relations cannot be expressed as analytic functions $\omega(k)$: $\mathrm{Re}D_i^{-1}(\omega, k, \Pi_i(\omega, k)) = 0$ and $\mathrm{Re}S_i^{-1}(\omega, k, \Sigma_i(\omega, k)) = 0$ are implicit functions for the dispersion relation since the self-energies cannot analytically be solved for ω. Instead, they have to be solved numerically and/or approximated. This implicit nature of the dependence of ω on k is indicated by the placement of k in the subscript instead of parentheses.

The results of numeric evaluations are shown in Figures 2.7 and 2.8. Due to the parity property (2.28) quarks and antiquarks obey identical dispersion relations up to the sign as do plasminos and antiplasminos.

2.6 The HTL grand canonical potential

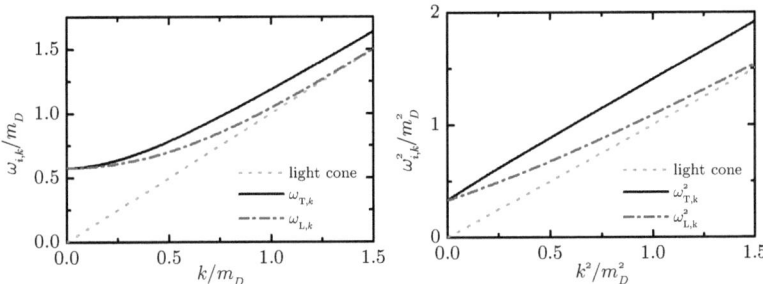

Figure 2.7: The dispersion relations $\omega_{\mathrm{T},k}$ of transverse and $\omega_{\mathrm{L},k}$ of longitudinal gluon modes scaled by the Debye mass are shown as functions of the momentum k scaled by the Debye mass in linear (left) and quadratic (right) scales.

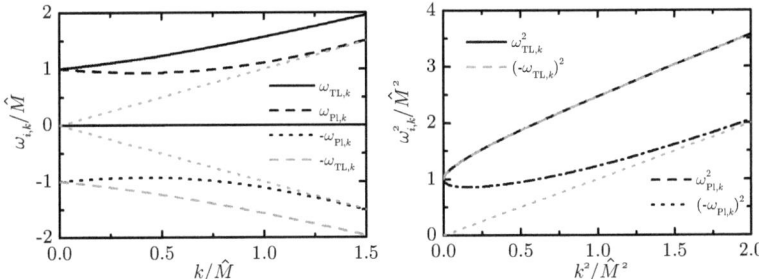

Figure 2.8: The dispersion relations $\omega_{i,k}$ of quarks (solid black curves), antiquarks (dashed grey curves), plasminos (black dashed curves) and antiplasminos (black dotted curves) scaled by the fermionic mass parameter are shown as functions of the momentum k scaled by the fermionic mass parameter in linear (left) and quadratic (right) scales. The dispersion relations of quarks and antiquarks as well as of plasminos and antiplasminos are equal up to the sign.

2.6 The HTL grand canonical potential

Given the explicit form of the HTL self-energies and the respective propagators, the remaining traces tr in Eq. (2.13) can be evaluated. Taking the trace in Minkowski space, the gluonic part decomposes into three contributions for one longitudinal and two (equivalent) transverse polarizations, while the quark contribution becomes the sum of the normal and the abnormal quark branch (positive and negative chirality over helicity ratio, respectively) when taking the Dirac trace. The remaining traces are simple, as they only give overall factors: the color trace $(N_c^2 - 1)$ for the gluons and N_c for the quarks, and the flavor and spin traces for the (light) quarks an additional $2N_l$. Consequently, we define the prefactors $d_g = N_c^2 - 1$ and $d_q = 2N_c N_l$.

For brevity we introduce the abbreviation $\int_{\mathrm{d}^4 k} = \int \mathrm{d}^4 k/(2\pi)^4$ and also omit the possible heavy quark flavor which can be added without difficulty at any stage as extra quark contribution with $\mu_s = 0$ and $d_s = 2N_c$ (cf. Section 2.3). The HTL grand canonical potential then reads (cf. [BIR01])

$$\begin{aligned}\frac{\Omega}{V} &= d_g \int_{d^4k} n_B \left\{ 2\mathrm{Im}\left(\ln D_T^{-1} - D_T \Pi_T\right) + \mathrm{Im}\left(\ln\left(-D_L^{-1}\right) + D_L \Pi_L\right) \right\} \\ &\quad + 2d_q \int_{d^4k} n_F \left\{ \mathrm{Im}\left(\ln S_+^{-1} - S_+ \Sigma_+\right) + \mathrm{Im}\left(\ln\left(-S_-^{-1}\right) + S_- \Sigma_-\right) \right\} - \frac{T}{V}\Gamma_2. \end{aligned} \qquad (2.29)$$

2.7 Effective coupling

It is clear that 2-loop QCD is only a crude approximation to the full theory. In order to accommodate further non-perturbative effects within the quasiparticle model, we introduce some flexibility by parameterizing the QCD coupling constant g^2 at vanishing chemical potential in a physically motivated way.

In consistency with the 1-loop approximation of the self-energies and propagators we utilize the renormalized coupling [PDG06] at 1-loop order

$$g^2(\bar{\mu}) = \frac{16\pi^2}{\beta_0 \ln(\bar{\mu}^2/\Lambda^2)}, \qquad (2.30)$$

where $\beta_0 = 11/3 - 2N_f/3$. It depends on the ratio of the renormalization scale $\bar{\mu}$ and the QCD scale parameter Λ. The former is usually related to the first Matsubara frequency $2\pi T$, while the latter represents the standard meter of the theory, i.e. a parameter to be determined by experiments. We determine the QCD scale by comparison with lattice calculations and absorb the 2π into a new QPM scale parameter $\lambda = \Lambda/2\pi$. We arrive at another valid parametrization, where the ratio $\bar{\mu}/\Lambda$ becomes T/λ.

In order to avoid the Landau pole of $g^2(T)$ at $T = \lambda$, the QCD coupling is substituted by an *effective coupling* $G^2(T)$ which is shifted by a temperature T_s

$$G^2(T, \mu = 0) = \frac{16\pi^2}{\beta_0 \ln\left(\frac{T-T_s}{\lambda}\right)^2}. \qquad (2.31)$$

With T_s appropriately chosen, G^2 behaves well within the plasma phase. Nevertheless it is still infrared (IR) divergent at some temperature within the hadronic phase. This can remedied by introducing a phenomenological IR regulator, e.g. a first [Blu04a] or higher [Sch07] order polynomial. Within the scope of the thesis however, we treat the QPM as being restricted to the area where its degrees of freedom constitute the elementary excitations.

The impact of including next-to-leading terms in the coupling have been subject of previous investigations [Sch07, Sch08b]. It was shown that including the truncated as well as the full 2-loop term provides little to no improvement but also no decline in the description of lattice data at vanishing chemical potential. The subtle changes of the coupling can be absorbed by a

2.8 Entropy density

reparametrization. Since this is the only place where it enters the model it is ample to use the 1-loop effective coupling.

At several points it is useful to use the scaled coupling $\alpha = G^2/4\pi$ in order to more transparently compare our results with perturbative calculations.

2.8 Entropy density

Differentiating the thermodynamic potential with respect to the temperature at constant chemical potential gives the entropy density as one of the state variables of the QGP. In contrast to the pressure, which is influenced by vacuum fluctuations, the entropy density is sensitive to thermal excitations only and therefore manifestly ultraviolet (UV) finite. Also, due to the stationarity of the thermodynamic potential under a variation with respect to the full propagators (cf. Eq. (2.8)), only explicit derivatives, i.e. derivatives of the statistical distribution functions within the thermodynamic potential, contribute to the derivative with respect to the temperature:

$$\frac{\partial \Omega}{\partial T} = \frac{\partial \Omega}{\partial T}\bigg|_{\text{expl.}} + \underbrace{\frac{\delta \Omega}{\delta D_i} \frac{\partial D_i}{\partial T}}_{0} + \underbrace{\frac{\delta \Omega}{\delta D_{0,i}} \frac{\partial D_{0,i}}{\partial T}}_{0}. \tag{2.32}$$

Therefore, the entropy density is ideally suited to investigate the properties of the QGP and is chosen to be the base quantity of the quasiparticle model. As a consequence, however, the pressure has to be reconstructed consistently introducing an additional integration constant (cf. Section 2.9).

Using $\text{Im}(D_T \Pi_T) = \text{Re} D_T \text{Im} \Pi_T + \text{Im} D_T \text{Re} \Pi_T$, the entropy density can be written as

$$s := -\frac{1}{V} \frac{\partial \Omega}{\partial T}\bigg|_{\mu} = s_{g,T} + s_{g,L} + s_{q,+} + s_{q,-} + s' \tag{2.33}$$

with

$$\begin{aligned}
s_{g,T} &= -2 d_g \int \frac{d^4 k}{(2\pi)^4} \frac{\partial n_B(\omega)}{\partial T} \left\{ \text{Im} \ln \left(+D_T^{-1}\right) - \text{Re} D_T \text{Im} \Pi_T \right\}, \\
s_{g,L} &= - d_g \int \frac{d^4 k}{(2\pi)^4} \frac{\partial n_B(\omega)}{\partial T} \left\{ \text{Im} \ln \left(-D_L^{-1}\right) + \text{Re} D_L \text{Im} \Pi_L \right\}, \\
s_{q,\pm} &= -2 d_q \int \frac{d^4 k}{(2\pi)^4} \frac{\partial n_F(\omega)}{\partial T} \left\{ \text{Im} \ln \left(\pm S_\pm^{-1}\right) \mp \text{Re} S_\pm \text{Im} \Sigma_\pm \right\}
\end{aligned} \tag{2.34}$$

and a residual entropy density s'. While each of the first four terms in (2.33) represents the entropy density of one particle species in the absence of the others, s' can be interpreted as the interaction entropy density between the different contributions. It contains terms of the form $\text{Im} D_T \text{Re} \Pi_T$ and the derivative of $\Gamma_2 T$ with respect to the temperature. At 2-loop order, these terms exactly cancel each other and thus $s' = 0$ [BIR01]. In fact, this seems to be a topological feature [CP75] which has been proven explicitly also for QED [VB98] and Φ^4 theory [Pes01a].

We now focus on the terms $\operatorname{Im}\ln(\pm D_{\mathrm{T,L}}^{-1})$ and $\operatorname{Im}\ln(\pm S_{\pm}^{-1})$, which equal the argument (i.e. the angle between the position vector representing the values in the complex plane and the positive real axis; cf. Appendix B.1) of the respective inverse propagators, and proceed by substituting the argument by the arc tangent (see also Appendix B.1), giving rise to an additional term compensating for its periodicity:

$$\begin{aligned}
\operatorname{Im}\left(\ln D_{\mathrm{T}}^{-1}\right) &= \arctan\left(\frac{\operatorname{Im}D_{\mathrm{T}}^{-1}}{\operatorname{Re}D_{\mathrm{T}}^{-1}}\right) + \pi\varepsilon(\operatorname{Im}D_{\mathrm{T}}^{-1})\Theta\left(-\operatorname{Re}D_{\mathrm{T}}^{-1}\right), \\
\operatorname{Im}\left(\ln\left(-D_{\mathrm{L}}^{-1}\right)\right) &= \arctan\left(\frac{\operatorname{Im}D_{\mathrm{L}}^{-1}}{\operatorname{Re}D_{\mathrm{L}}^{-1}}\right) - \pi\varepsilon(\operatorname{Im}D_{\mathrm{L}}^{-1})\Theta\left(+\operatorname{Re}D_{\mathrm{L}}^{-1}\right), \\
\operatorname{Im}\left(\ln S_{+}^{-1}\right) &= \arctan\left(\frac{\operatorname{Im}S_{+}^{-1}}{\operatorname{Re}S_{+}^{-1}}\right) + \pi\varepsilon(\operatorname{Im}S_{+}^{-1})\Theta\left(-\operatorname{Re}S_{+}^{-1}\right), \\
\operatorname{Im}\left(\ln\left(-S_{-}^{-1}\right)\right) &= \arctan\left(\frac{\operatorname{Im}S_{-}^{-1}}{\operatorname{Re}S_{-}^{-1}}\right) - \pi\varepsilon(\operatorname{Im}S_{-}^{-1})\Theta\left(+\operatorname{Re}S_{-}^{-1}\right). \quad (2.35)
\end{aligned}$$

From the properties of the imaginary parts of the self-energies (2.27), we find $\varepsilon(\operatorname{Im}D_i^{-1}(\omega)) = -\varepsilon(\omega)$ for the gluons and $\varepsilon(\operatorname{Im}S_{\pm}(\omega)) \equiv -1$ for the normal and abnormal quark branches. We end up with

$$\begin{aligned}
s_{g,\mathrm{T}} &= +2d_g\int_{\mathrm{d}^4 k}\frac{\partial n_{\mathrm{B}}}{\partial T}\left\{\pi\varepsilon(\omega)\Theta\left(-\operatorname{Re}D_{\mathrm{T}}^{-1}\right) - \arctan\frac{\operatorname{Im}\Pi_{\mathrm{T}}}{\operatorname{Re}D_{\mathrm{T}}^{-1}} + \operatorname{Re}D_{\mathrm{T}}\operatorname{Im}\Pi_{\mathrm{T}}\right\}, \\
s_{g,\mathrm{L}} &= -d_g\int_{\mathrm{d}^4 k}\frac{\partial n_{\mathrm{B}}}{\partial T}\left\{\pi\varepsilon(\omega)\Theta\left(+\operatorname{Re}D_{\mathrm{L}}^{-1}\right) - \arctan\frac{\operatorname{Im}\Pi_{\mathrm{L}}}{\operatorname{Re}D_{\mathrm{L}}^{-1}} + \operatorname{Re}D_{\mathrm{L}}\operatorname{Im}\Pi_{\mathrm{L}}\right\}, \\
s_{q,\pm} &= \pm 2d_q\int_{\mathrm{d}^4 k}\frac{\partial n_{\mathrm{F}}}{\partial T}\left\{\pi\Theta\left(\mp\operatorname{Re}S_{\pm}^{-1}\right) - \arctan\frac{\operatorname{Im}\Sigma_{\pm}}{\operatorname{Re}S_{\pm}^{-1}} + \operatorname{Re}S_{\pm}\operatorname{Im}\Sigma_{\pm}\right\}.
\end{aligned}$$
$$(2.36)$$

The partial entropy densities (2.36) are and therefore the whole entropy density expression (2.33) is independent of possible renormalization factors. As required, the expression is also explicitly UV finite, as the derivatives of the distribution functions tame the UV behavior.

The quark entropy density $s_q = s_{q,+} + s_{q,-}$ can be simplified by utilizing the parity properties for quark propagators (2.28) and self-energies (2.26). Introducing the distribution function of antiparticles

$$n_{\mathrm{F}}^A = \frac{1}{e^{\beta(\omega+\mu)}+1} \quad (2.37)$$

with

$$\frac{\partial n_{\mathrm{F}}(-\omega)}{\partial\omega} = -\frac{\partial n_{\mathrm{F}}^A(\omega)}{\partial\omega} \quad (2.38)$$

and substituting $\omega \to -\omega$ within $s_{q,-}$, we find

$$s_q = 2d_q\int_{\mathrm{d}^4 k}\left(\frac{\partial n_{\mathrm{F}}}{\partial T}+\frac{\partial n_{\mathrm{F}}^A}{\partial T}\right)\left\{\pi\Theta(-\operatorname{Re}S_{+}^{-1}) - \arctan\left(\frac{\operatorname{Im}\Sigma_{+}}{\operatorname{Re}S_{+}^{-1}}\right) + \operatorname{Re}S_{+}\operatorname{Im}\Sigma_{+}\right\}.$$
$$(2.39)$$

2.8 Entropy density

In the rearranged form, the quasiparticle contributions from the pole term $\pi\Theta(\text{-Re}S_+^{-1})$ are now more clearly identified. While the contributions of the plasminos to the entropy density are given by the energy integration from $-\infty$ to 0, the integration from 0 to $+\infty$ yields the contributions of the quarks (in both cases including the respective antiparticles). Isolating both parts of the spectrum by applying the parity properties in the negative energy domain once more gives the explicit expressions

$$s_{q,\text{TL}} = 2d_q \int_{d^3k} \int_0^\infty \frac{d\omega}{2\pi}\,()\,\left\{\pi\Theta(\text{-Re}S_+^{-1}) - \arctan\left(\frac{\text{Im}\Sigma_+}{\text{Re}S_+^{-1}}\right) + \text{Re}S_+\text{Im}\Sigma_+\right\},$$

$$s_{q,\text{Pl}} = -2d_q \int_{d^3k} \int_0^\infty \frac{d\omega}{2\pi}\,()\,\left\{\pi\Theta(\text{Re}S_-^{-1}) - \arctan\left(\frac{\text{Im}\Sigma_-}{\text{Re}S_-^{-1}}\right) + \text{Re}S_-\text{Im}\Sigma_-\right\},$$

(2.40)

where the sum of the derivatives of the distribution functions is abbreviated by the round parentheses (). While this separation seems straightforward, it has to be handled with care as the Landau damping term within the quark self-energies $\Sigma_\pm$ (see the imaginary parts in Fig. 2.4) can in general not be separated into quark and plasmino contributions in this simple way.

For the following, it is a good choice to also rewrite the gluon expression with the energy integral over positive ω only. To this end, we introduce the general abbreviation of the curly brackets $\{\}_i$, where the index i denotes the quasiparticle family. From the symmetric self-energies we find antisymmetric brackets $\{\}_{g,\text{T/L}}$ in Eqs. (2.36) which together with the antisymmetric derivative of the distribution function leads to symmetric integrand. Thus, we have

$$s_{g,\text{T}} = 4d_g \int_{d^3k} \int_0^\infty \frac{d\omega}{2\pi} \frac{\partial n_\text{B}}{\partial T} \{\}_{g,\text{T}},$$

$$s_{g,\text{L}} = -2d_g \int_{d^3k} \int_0^\infty \frac{d\omega}{2\pi} \frac{\partial n_\text{B}}{\partial T} \{\}_{g,\text{L}}.$$

(2.41)

The striking feature of the entropy expressions derived from the 1-loop HTL self-energies expresses itself in the absence of interaction terms. Therefore, it is justified to speak of quasiparticles. The interaction of real quarks and gluons is encoded within the modified properties of these quasiparticles (e.g. self-energies/dispersion relations), their damping behavior and the plasma modes quantized by the collective excitations.

2.9 The pressure

As our model is based on the entropy density we need to self-consistently reconstruct the pressure as thermodynamic potential. Since the entropy density is required to be stationary with respect to a variation of the propagators (or – via the dispersion relation – equivalently the self-energies; cf. Eq. (2.32)) only explicit derivatives must contribute to the derived quantities entropy density and net quark density.

However, if using the natural definition for the partial pressures by replacing the derivatives of the distribution functions in the partial entropy densities s_i (Eqs. (2.40) and (2.41)[11]) with the distribution functions themselves (cf. also [Sch09])

$$p_{g,\mathrm{T}} = +4d_g \int_{\mathrm{d}^3k} \int_0^\infty \frac{\mathrm{d}\omega}{2\pi} \, n_\mathrm{B} \{\}_{g,\mathrm{T}},$$

$$p_{g,\mathrm{L}} = -2d_g \int_{\mathrm{d}^3k} \int_0^\infty \frac{\mathrm{d}\omega}{2\pi} \, n_\mathrm{B} \{\}_{g,\mathrm{L}},$$

$$p_{q,\mathrm{TL}} = 2d_q \int_{\mathrm{d}^3k} \int_0^\infty \frac{\mathrm{d}\omega}{2\pi} \left(n_\mathrm{F} + n_\mathrm{F}^A\right) \{\}_{q,+},$$

$$p_{q,\mathrm{Pl}} = -2d_q \int_{\mathrm{d}^3k} \int_0^\infty \frac{\mathrm{d}\omega}{2\pi} \left(n_\mathrm{F} + n_\mathrm{F}^A\right) \{\}_{q,-}, \qquad (2.42)$$

we find that implicit derivatives contribute via the self-energies

$$\frac{\partial p_i}{\partial T} = \left.\frac{\partial p_i}{\partial T}\right|_{\mathrm{expl.}} + \frac{\partial p_i}{\partial \Pi_i}\frac{\partial \Pi_i}{\partial T},$$

$$\frac{\partial p_i}{\partial \mu} = \left.\frac{\partial p_i}{\partial \mu}\right|_{\mathrm{expl.}} + \frac{\partial p_i}{\partial \Pi_i}\frac{\partial \Pi_i}{\partial \mu}. \qquad (2.43)$$

The most prudent way[12] to deal with these contributions is to define the overall pressure p as

$$p := \sum p_i - B, \qquad (2.44)$$

where $B = \sum B_i$ is a mean field pressure contribution, resembling to a bag pressure, which here

---

[11] For convenience, entropy density expressions with energy integrals $\omega \in [0,\infty)$ are used to avoid any infinite integration constants of the form $\int_{-\infty}^0 \mathrm{d}\omega\, n_i()$ with $n_i(\omega \to -\infty) \neq 0$. If occurring they have to be regulated and are cancelled by the overall integration constant B_c (cf. Section 2.11). Due to the latter being fixed to lattice calculations the integration constants of the partial pressures are ultimately irrelevant and, as a convention, set to 0.

[12] For an insightful review of alternative choices, cf. [Gar09].

2.9 The pressure

is defined by its derivatives

$$\frac{\partial B}{\partial T} := \sum \frac{\partial p_i}{\partial \Pi_i} \frac{\partial \Pi_i}{\partial T} \quad \text{and} \quad \frac{\partial B}{\partial \mu} := \sum \frac{\partial p_i}{\partial \Pi_i} \frac{\partial \Pi_i}{\partial \mu}, \tag{2.45}$$

in such a way that indeed entropy density

$$s := \frac{\partial p}{\partial T} = \sum \left. \frac{\partial p_i}{\partial T} \right|_{\text{expl.}} + \frac{\partial p_i}{\partial \Pi_i} \frac{\partial \Pi_i}{\partial T} - \sum \frac{\partial p_i}{\partial \Pi_i} \frac{\partial \Pi_i}{\partial T} = \sum \left. \frac{\partial p_i}{\partial T} \right|_{\text{expl.}} \tag{2.46}$$

and – yet unspecified – net quark density

$$n := \frac{\partial p}{\partial \mu} = \sum \left. \frac{\partial p_i}{\partial \mu} \right|_{\text{expl.}} + \frac{\partial p_i}{\partial \Pi_i} \frac{\partial \Pi_i}{\partial \mu} - \sum \frac{\partial p_i}{\partial \Pi_i} \frac{\partial \Pi_i}{\partial \mu} = \sum \left. \frac{\partial p_i}{\partial \mu} \right|_{\text{expl.}} \tag{2.47}$$

follow from the expressions (2.42) with the explicit derivatives w.r.t. temperature and chemical potential, respectively, acting on the distribution functions only.

For the overall pressure p defined in this way to be a potential, the condition

$$\frac{\partial^2 p}{\partial T \partial \mu} = \frac{\partial^2 p}{\partial \mu \partial T} \tag{2.48}$$

has to be obeyed. In the framework of thermodynamics such a requirement is referred to as Maxwell's relation. While it is true for the partial pressures p_i above due to the theorem of Schwarz, it has to be ensured for the mean field pressure B, since the latter is per definitionem not necessarily an analytic function.

From Eqs. (2.45) we find

$$\begin{aligned}\frac{\partial^2 B}{\partial \mu \partial T} &= \sum \frac{\partial^2 p_i}{\partial \mu|_{\text{ex.}} \partial \Pi_i} \frac{\partial \Pi_i}{\partial T} + \frac{\partial^2 p_i}{(\partial \Pi_i)^2} \frac{\partial \Pi_i}{\partial \mu} \frac{\partial \Pi_i}{\partial T} + \frac{\partial p_i}{\partial \Pi_i} \frac{\partial^2 \Pi_i}{\partial \mu \partial T}, \\ \frac{\partial^2 B}{\partial T \partial \mu} &= \sum \frac{\partial^2 p_i}{\partial T|_{\text{ex.}} \partial \Pi_i} \frac{\partial \Pi_i}{\partial \mu} + \frac{\partial^2 p_i}{(\partial \Pi_i)^2} \frac{\partial \Pi_i}{\partial T} \frac{\partial \Pi_i}{\partial \mu} + \frac{\partial p_i}{\partial \Pi_i} \frac{\partial^2 \Pi_i}{\partial T \partial \mu},\end{aligned} \tag{2.49}$$

where the the middle terms are trivially equal. Also, due to Schwarz's theorem, the respective third terms match and the derivatives of p_i in the first term can be interchanged. Therefore, matching the derivatives of p_i with s_i and n_i, the integrability condition for B is

$$\sum \frac{\partial s_i}{\partial \Pi_i} \frac{\partial \Pi_i}{\partial \mu} = \sum \frac{\partial n_i}{\partial \Pi_i} \frac{\partial \Pi_i}{\partial T} \tag{2.50}$$

which is commonly dubbed *flow equation*. It presents – of course – the same condition which follows from requiring Eq. (2.48) directly for model consistency, using Eqs. (2.46, 2.47) and dismissing the explicit terms due to the theorem of Schwarz.

Thus, the flow equation follows naturally as both, integrability condition for the mean field pressure B and consistency condition of the model. The striking feature about it is, however, that it relates information known, e.g. about the net quark density dependence on the temperature, to the hitherto unknown dependence of the entropy density on the chemical potential, thus allowing for an extrapolation in the direction of the chemical potential. The mean field pressure is essential to ensure the consistency of this procedure.

It is important to note that it is not possible to construct separate thermodynamic potentials

$$p_{(i)} := p_i - B_i \qquad (2.51)$$

from the single contributions B_i to the overall mean field pressure

$$B = B_c + \sum B_i \qquad (2.52)$$

as thermodynamic consistency is ensured by the flow equation for the overall pressure $p = \sum p_i - B$ only. For instance $\partial^2 B_i/\partial T \partial \mu = \partial^2 B_i/\partial \mu \partial T$ does not hold for the separate contributions B_i making them – as well as the $p_{(i)}$ – path-dependent. As a result, all B_i for one (T, μ) have to be computed along the same path and only the overall mean field pressure $B(T, \mu)$ can be compared between different paths (e.g. different characteristic curves of the flow equation). Thus, most of the thermodynamic quantities following from and including the pressure such as the the energy density or the interaction measure also have path-dependent partial contributions. Since the consistency of the overall quantities is ensured the deviations are shifts among the individual contributions.

In the QPM, entropy density and net quark density are distinct quantities as they are known by analytic expressions depending on the effective coupling G^2 only, but not on the pressure and any integration paths. Thus, regarding the characteristics not as paths in but a grid on the T-μ-plane where a unique $G^2(T, \mu)$ exists, they are manifestly path independent. Still, since the flow equation (2.50) only ensures the consistency of the overall net quark and entropy density it is clear that, in general, consistency of the partial contributions $\partial n_i/\partial T = \partial s_i/\partial \mu$ is not provided for. This is, of course, nothing more but a different way expressing that the $p_{(i)}$ are not thermodynamic potentials. As a result, the individual contributions to thermodynamic quantities – path-dependent or not – can only be taken as indications of the size of such a contributions as consistency among these is not ensured.

While it is straightforward to combine the quark entropy density contribution $s_{q,\pm}$ (2.36) or $s_{q,\mathrm{TL}}$ and $s_{q,\mathrm{Pl}}$ (2.40) into one expression (2.39) by including the respective counterpart via the negative energy domain and the parity properties, this is not as easy for the pressure due to the combined distribution functions $(n_\mathrm{F} + n_\mathrm{F}^A)$ not being symmetric[13] at the point of origin.

We close this Section by giving the explicit expression for the net particle densities. They

---

[13]Naively employing a combined expression leads to a divergence as $(n_\mathrm{F} + n_\mathrm{F}^A) \to 2$ for $\omega \to -\infty$ while vanishing for $\omega \to \infty$. While this can be cured by the introduction of an (infinite) integration constant $-\int_{-\infty}^{0} 2$ the resulting expression $(n_\mathrm{F} + n_\mathrm{F}^A - 2\Theta(-\omega))$ for the combined distribution function does not contribute to a more elegant and insightful formulation and is, therefore, not used.

are easily obtained by replacing the derivatives of the distribution functions in the entropy density expressions with respect to the temperature by derivatives with respect to the chemical potential, as prescribed by Eqs. (2.46) and (2.47). For the gluons as bosons and strange quarks as fermions with vanishing chemical potential, the net particle density is zero. For the light quarks we find, by replacing the derivatives of the distribution functions with respect to the temperature in Eq. (2.39) with their equivalents with respect to the chemical potential,

$$n_q = 2d_q \int_{\mathrm{d}^4 k} \left(\frac{\partial n_\mathrm{F}}{\partial \mu} + \frac{\partial n_\mathrm{F}^A}{\partial \mu} \right) \left\{ \pi\Theta(-\mathrm{Re}S_+^{-1}) - \arctan\left(\frac{\mathrm{Im}\Sigma_+}{\mathrm{Re}S_+^{-1}}\right) + \mathrm{Re}S_+ \mathrm{Im}\Sigma_+ \right\}. \qquad (2.53)$$

As for the entropy density, this expression can be split into a particle and a plasmino contribution. The expressions resemble those of Eqs. (2.40) with the replacement of the distribution functions in the parentheses ().

2.10 Solution of the flow equation

While it is possible to employ the flow equation directly on entropy density and net quark density, it proves more fruitful to apply it on a more fundamental level. Entropy density and net quark density, as well as all other state variables, depend on the effective coupling G^2 via the quasiparticle self-energies. Thus, on its basic level, the flow equation links information about the effective coupling in the direction of temperature and chemical potential.

Writing the derivatives of the self-energies with respect to T and μ in terms of explicit and implicit (via G^2) derivatives

$$\begin{aligned}
\frac{\partial \Pi_i}{\partial T} &= \left.\frac{\partial \Pi_i}{\partial T}\right|_{G^2} + \frac{\partial \Pi_i}{\partial G^2}\frac{\partial G^2}{\partial T}, \\
\frac{\partial \Pi_i}{\partial \mu} &= \left.\frac{\partial \Pi_i}{\partial \mu}\right|_{G^2} + \frac{\partial \Pi_i}{\partial G^2}\frac{\partial G^2}{\partial \mu}
\end{aligned} \qquad (2.54)$$

and reordering the flow equation

$$\underbrace{-\sum \frac{\partial n_i}{\partial \Pi_i}\frac{\partial \Pi_i}{\partial G^2}\frac{\partial G^2}{\partial T}}_{a_T} + \underbrace{\sum \frac{\partial s_i}{\partial \Pi_i}\frac{\partial \Pi_i}{\partial G^2}\frac{\partial G^2}{\partial \mu}}_{a_\mu} = \underbrace{\sum \frac{\partial n_i}{\partial \Pi_i}\left.\frac{\partial \Pi_i}{\partial T}\right|_{G^2} - \frac{\partial s_i}{\partial \Pi_i}\left.\frac{\partial \Pi_i}{\partial \mu}\right|_{G^2}}_{b} \qquad (2.55)$$

leads to the elliptic quasilinear partial differential equation (PDE)

$$a_T \frac{\partial G^2}{\partial T} + a_\mu \frac{\partial G^2}{\partial \mu} = b \qquad (2.56)$$

for the effective coupling G^2. In this form, the flow equation is known as Peshier equation [Pes00].

The flow equation closely resembles a one-dimensional transport equation (hence the name) and similarly can be solved by conversion to a system of coupled ordinary differential equations using the method of characteristics. Introducing a curve parameter x, so that $T = T(x)$, $\mu = \mu(x)$ and $G^2 = G^2(T(x), \mu(x))$, we have

$$\frac{\partial G^2}{\partial x} = \frac{\partial G^2}{\partial T}\frac{\partial T}{\partial x} + \frac{\partial G^2}{\partial \mu}\frac{\partial \mu}{\partial x}. \tag{2.57}$$

Comparison with the flow equation (2.56) then gives

$$a_T = -\frac{\partial T}{\partial x}, \quad a_\mu = -\frac{\partial \mu}{\partial x} \quad \text{and} \quad b = -\frac{\partial G^2}{\partial x}. \tag{2.58}$$

The system of coupled ordinary differential equations given by $y' := (T, \mu, G^2)' = (-a_T, -a_\mu, -b) = f(x, y)$ together with the boundary conditions $y(x = 0) = (T_0, 0, G^2(T_0, \mu = 0))$ represents a Cauchy problem, where the effective coupling $G^2(T_0, \mu = 0)$ on the boundary is fixed in such a way that the QPM describes certain lattice quantities. The existence and uniqueness of the solution is ensured by the Picard–Lindelöf theorem as the Lipschitz condition is guaranteed by the fact that a_T, a_μ and b are continuously differentiable with respect to T, μ and G^2.

It is solved using standard numerical techniques (adaptive stepsize solver LSODA [Hin83], classical Runge-Kutta method for verification). The result of any solution with boundary value $(T_0, 0, G^2(T_0, \mu = 0))$ is a characteristic curve $(T(x), \mu(x))$ emerging at $(T_0, \mu = 0)$ and ending at $(T = 0, \mu_f)$ where $\mu_f := \mu(x_f)$ with $x_f := x(T = 0)$. Along the characteristic curve $G^2(T(x), \mu(x))$ is given.

From the general thermodynamic properties for the state variables $n(\mu \to 0) \longrightarrow 0$ and $s(T \to 0) \longrightarrow 0$ (Nernst's law) and thus also for the derivatives with respect to the self-energies follow $a_T(\mu \to 0) \longrightarrow 0$ and $a_\mu(T \to 0) \longrightarrow 0$. Hence, the characteristics approach the T and the μ axis perpendicularly. The signs in (2.58) are chosen deliberately so that the characteristics evolve in the direction of increasing chemical potential and, as a consequence, decreasing temperature ($a_\mu < 0$ and $a_T > 0$). It can also be shown that $b(T, \mu \to 0) \longrightarrow +0$, implying that the coupling G^2 increases for small chemical potentials along the characteristics, and – taking into account $a_T(\mu \to 0) \longrightarrow 0$ – has a local minimum with respect to the chemical potential at $\mu = 0$.[14]

From $\partial T/\partial \mu = a_T/a_\mu$ it is clear that the path of the characteristics is governed by the ratio of the two coefficients. An estimate for the relation between emergence temperature T_0 and incidence chemical potential μ_f of a characteristic can be given by the average ratio of coefficients as

$$\frac{\int dx\, a_T}{\int dx\, a_\mu} = \frac{\int_0^{x_f} dx \frac{\partial T}{\partial x}}{\int_0^{x_f} dx \frac{\partial \mu}{\partial x}} = \frac{-T_0}{\mu_f} \tag{2.59}$$

and thus

$$\frac{T_0}{\mu_f} \approx -\frac{\overline{a_T}}{\overline{a_\mu}} \approx -\frac{a_T(T = 0)}{a_\mu(\mu = 0)}, \tag{2.60}$$

---

[14]From the flow equation we have $\partial G^2/\partial \mu|_{\mu=0} = (b/a_\mu - (a_T/a_\mu)(\partial G^2/\partial T))|_{\mu=0} = 0$.

2.10 Solution of the flow equation

where in the last step, under the assumption of approximately linear evolution of a_T and a_μ, the fact that $a_T(\mu = 0) = a_\mu(T = 0) = 0$ is exploited. It is important to realize that the paths of the characteristics do not necessarily have any influence on the behavior of the extrapolated quantity (the effective coupling G^2). However, there are some quantities which can be shown to be (almost) constant along the characteristic curves (cf. Section 4.7).

The characteristic curves do not necessarily cover the whole domain of the PDE. For the quasiparticle model, a large region at $T < T_c$ and small chemical potential is inaccessible to a numerical solution. However, this numerical deficiency turns out to provide an important insight into the limits of the description using our (quasi)quark and gluon degrees of freedom. The region unreachable by characteristic curves will have to described using a different approach. Then again, this does not necessarily imply that, in turn, the quasiparticle method is valid for all regions accessible to characteristics. Physical justifications are necessary to determine the limits of applicability of a certain approach. One such limit is lies in the fact that, if the pressure turns negative along a characteristic curve, all other states of matter are preferred and a phase transition is inevitable. Keeping this in mind, it is worth mentioning that one may actually find solutions for regions inaccessible to characteristics in a weak (integral) sense. In wave theory they turn out to be rarefaction waves leading to the term rarefactions for the affected areas [Kno00].

Besides rarefactions, regions of crossing characteristic curves may appear in the solution of the PDE as a sign of impaired self-consistency or inconsistent initial conditions. The size of these regions is related to the deviance from self-consistency and may serve as a measure of the latter. Crossing characteristics are, however, not generally insurmountable barriers and, in fact, can be dealt with if handled correctly [Kno00]. For instance in wave theory the crossings are – opposed to the rarefaction waves – closely related to shock waves.

In essence, the treatment of crossing characteristics amounts to an affirmation of the underlying conservation laws. While generally the validity of the model predictions may be in question for all multivalued solutions, this criterion provides the unique physical trajectory. In many cases the latter may only be piecewise smooth with one or more jump discontinuities, however the case is more simple for thermodynamic models, where conservation laws are explicitly obeyed and thermodynamic quantities such as the energy or entropy density are continuous functions. Taking the set of characteristic curves from the regular region together with the maximum set of neighboring, non-crossing characteristic within the region of multivalued solutions lines yields continuous and unique-valued state variables as functions of the thermodynamic parameters covered by those lines. The state variables transit smoothly into the region of multivalued solutions thereby obeying all conservation laws. Any other choice of characteristics/solutions leads to severe discontinuities in all state variables so that, in general, the conservation of all relevant quantities is out of the question.

2.11 Integration of the mean field pressure

The quantity which ensures thermodynamic self-consistency is the mean field pressure (cf. Section 2.9). It is known by its derivatives with respect to temperature and chemical potential only. It can be integrated straightforward by

$$B = B_c + \int dT \frac{\partial B}{\partial T} + \int d\mu \frac{\partial B}{\partial \mu}, \qquad (2.61)$$

where B_c acts as overall pressure integration constant, eliminating the necessity of partial pressure integration constants in Eqs. (2.42).

This method, however, requires the derivative of the effective coupling with respect to the chemical potential and the pressure, which is not known along one characteristic curve[15]. Therefore, it is more convenient to perform the integration along the characteristic itself, using the knowledge about $\partial G^2/\partial x = b$. The elements of the Jacobian along these paths are the negative coefficients of the flow equation (cf. Eqs. (2.58)). Using the definitions of the derivatives of B (cf. Eqs. (2.45)) we have

$$B = B_0 - \int dx \sum_i \left(a_T \left.\frac{\partial \Pi_i}{\partial T}\right|_{G^2} + a_\mu \left.\frac{\partial \Pi_i}{\partial \mu}\right|_{G^2} + b \frac{\partial \Pi_i}{\partial G^2} \right) \frac{\partial p_i}{\partial \Pi_i}, \qquad (2.62)$$

where $B_0 = B(T_0)$ is the mean field pressure at the initial temperature T_0 of the characteristic curve at vanishing chemical potential – obtained from Eq. (2.61) by integrating along the T-axis. For details see Appendix D.[16]

At finite chemical potential, it is actually not inevitably necessary to compute the mean field pressure. As thermodynamic self-consistency of the overall pressure is ensured by the flow equation it suffices to integrate the pressure from entropy density and net quark density. From

$$p = p_0 + \int dT \frac{\partial p}{\partial T} + \int d\mu \frac{\partial p}{\partial \mu} \qquad (2.64)$$

we find

$$p = p_0 + \int dT \sum s_i + \int d\mu \sum n_i \qquad (2.65)$$

---

[15] Essentially, the mean field pressure may be obtained according to Eq. (2.61) by integrating alternately with respect to the chemical potential and with respect to the temperature between two characteristic curves (approximating the characteristic curves by a stairway). The derivatives of the effective coupling are available by comparison of the effective couplings along both characteristics. With decreasing distance of the characteristic curves, the result equals the standard method.

[16] In order to verify the consistency of the numerical implementation it is useful to check if

$$\frac{\partial B}{\partial T} = \frac{\partial p}{\partial T} - s \quad \text{and} \quad \frac{\partial B}{\partial \mu} = \frac{\partial p}{\partial \mu} - n, \qquad (2.63)$$

as required by Eqs. (2.43) and the definitions of B, n and s.

or along a characteristic curve

$$p = p_0 + \int \mathrm{d}x \left(a_T s_i + a_\mu n_i\right),\qquad(2.66)$$

where $p_0 = p(T_0, \mu = 0)$ with T_0 being the emergence temperature of the characteristic. In order to keep the calculation of the pressure at finite chemical potential consistent with the evaluation at vanishing μ via Eq. (2.44) we prefer the former method using the mean field pressure. The latter one has been used to cross-check the numerical results.

2.12 Interaction measure – connection to lattice QCD

In order to fix our model parameters, i.e. the parameters λ and T_s of the effective coupling and the pressure integration constant B_c, we compare our model to first-principle computations of QCD bulk properties on the lattice (cf. Section 1.2). The primary output of a lattice calculation is the trace anomaly in units of the fourth power of the temperature $T^{\mu\mu}/T^4 = \Delta/T^4 = (e - 3p)/T^4$, also dubbed interaction measure [Baz09] (recall that $T^{\mu\nu}$ is the energy-momentum tensor, cf. Eqs. (1.4)).

At vanishing chemical potential, all other state variables then follow from Δ by integration and general thermodynamic relations:

$$\begin{aligned}\frac{p(T)}{T^4} &= \frac{p(T_c)}{T_c^4} + \int_{T_c}^{T} dT' \frac{\Delta(T')}{T'^5}, \\ \frac{s}{T^3} &= 4\frac{p}{T^4} + \frac{\Delta}{T^4}, \\ \frac{e}{T^4} &= 3\frac{p}{T^4} + \frac{\Delta}{T^4}.\end{aligned}\qquad(2.67)$$

The QCD interaction measure is a very characteristic quantity displaying a peak structure, the maximum of which arises from a turning point of the scaled pressure. From general thermodynamic relations the temperature T_p of the maximum follows from the pressure and its derivatives via

$$\left(\frac{16}{T^4} - \frac{7}{T^3}\frac{\partial}{\partial T} + \frac{1}{T^2}\frac{\partial^2}{(\partial T)^2}\right) p \bigg|_{T=T_p} = 0.\qquad(2.68)$$

Within our quasiparticle model, the location of the maximum is governed by the values of the parameters T_s and λ, where the latter one also affects the peak width. The peak height of Δ/T^4 is essentially determined by B_0.

As opposed to the QPM, a pressure constant $p(T_c)$ has to be specified for all thermodynamic variables. The chosen value of the T_c should be at the lower limit of our model for deconfinement, i.e. $T_c \approx 190$ MeV according to [Baz09] and $T_c \approx 152$ MeV according to [Bor10a, Bor10b]. The values of $p(T_c)$ are determined by integration of the interaction measure below the phase transition via the first of Eqs. (2.67) and/or comparison to the pressure of hadron models such as the hadron resonance gas (HRG). In essence, $p(T_c)$ contains information from the hadronic phase.

The temperature T_c at $\mu = 0$ is commonly dubbed pseudocritical temperature, referring to the crossover between the hadronic and the quark-gluon degrees of freedom at vanishing chemical potential. The characteristic emerging from the pseudocritical temperature may be taken as a possible guide for the presumed transition line (pseudocritical line) between the two phases [KBS06] and is therefore called pseudocritical characteristic.

In contrast to the lattice case, the QPM is based on the entropy density, i.e. pressure and interaction measure are analytic integrals of the entropy including an integration constant B_c. Thus, fitting solely the QPM interaction measure to the lattice interaction measure allows not only to determine the parameters T_s and λ but also the pressure integration constant B_c. Thus, the QPM pressure as well as entropy and energy density follow directly from $e - 3p$ without another integration constant. This is due to the additional knowledge of explicit expressions for all thermodynamic quantities. Within the model, $p(T_c)$ is known from the parameters T_s, λ and B_c via the expressions of s and Δ in $p(T_c) = (Ts(T_c; T_s, \lambda) - \Delta(T_c; T_s, \lambda, B_c))/4$.

This naive fit is, however, problematic considering the information conveyed within the integration constants.[17] While $p(T_c)$ from the lattice depicts information about the interaction measure below T_c, the pressure integration constant B_c found by comparing to the interaction measure of QPM and lattice contains information from above T_c only. Thus, B_c is only an additional degree of freedom in the fit procedure which – due to thermodynamic consistency – describes the pressure to a certain extent, however, entirely neglects the additional information from the lattice about the hadronic phase.

Therefore, it is prudent to use a different adjustment procedure for the QPM parameters. Instead of fixing B_c in the fit to the interaction measure, the parameter is is varied at any given (λ, T_s) so that $p_{\mathrm{QPM}}(T_c) = p_{\mathrm{lattice}}(T_c)$. Thus, by requiring an exact translation of the pressure at T_c from the lattice calculations into the QPM, the model fit is reduced to the two parameters of the effective coupling. In a certain sense, this method allows to include information about the interaction measure of the confined phase even though the model is valid for the degrees of freedom of the deconfined phase only.

---

[17]Additionally, in most cases the pressure integration constant obtained by a solely fitting to the interaction measure does only provide a satisfactory description of the pressure. This is due to the (known statistical and unknown systematic) errors of the lattice results. A fit using a certain value B_c might, together with modified λ and T_s, provide a better description of the lattice interaction measure data while the description of the pressure is only average, than a set (λ, T_s, B_c) describing both equally well.

3 Analytic investigation of the model

3.1 Asymptotic dispersion relations and quark restmasses

In order to acquire simple analytic expressions for the thermodynamic variables it is sometimes necessary to have explicit yet approximated dispersion relations $\omega_i(k)$. A prudent choice is to use the asymptotic dispersion relations near the light cone as a reasonable simplification of the full HTL dispersion relations. For transverse gluons, where $0 = \mathrm{Re}D_\mathrm{T}^{-1} = -\omega_{\mathrm{T},k}^2 + k^2 + \mathrm{Re}\Pi_\mathrm{T}(\omega_{\mathrm{T},k}, k)$, they can be obtained by a first order iterative approximation

$$\omega_{\mathrm{T},k}^2 = k^2 + \mathrm{Re}\Pi_\mathrm{T}(\omega_{\mathrm{T},k}, k) \approx k^2 + \mathrm{Re}\Pi_\mathrm{T}(k, k) \tag{3.1}$$

giving an explicit dispersion relation of the form

$$\omega_\mathrm{T}^2(k) = k^2 + m_{g,\infty}^2 \tag{3.2}$$

with the asymptotic gluon mass

$$m_{g,\infty}^2 := \mathrm{Re}\Pi_\mathrm{T}(k,k) = m_D^2/2 \tag{3.3}$$

which is independent of both energy and momentum.

In order to derive the asymptotic dispersion relation for quarks from $\mathrm{Re}S_+^{-1} = -\omega + k + \mathrm{Re}\Sigma_+(\omega, k) \stackrel{!}{=} 0$ we multiply by k and complete the square

$$0 = -\omega k + k^2 + \mathrm{Re}\Sigma_+ k + \omega^2 - \omega^2 + \omega k - \omega k$$

which leads to

$$\omega^2 = (\omega - k)^2 + \mathrm{Re}\Sigma_+ k + (k + \mathrm{Re}\Sigma_+)k.$$

The difference $(\omega - k)^2$ can be neglected near the light cone and $\mathrm{Re}\Sigma_+(\omega, k)$ may be approximated by its first order iteration $\mathrm{Re}\Sigma_+(k, k)$ as well. If ω is positive, the result

$$\omega_\mathrm{TL}^2(k) = k^2 + m_{q,\infty}^2 \tag{3.4}$$

is the asymptotic dispersion relation for quarks with the asymptotic quark mass

$$m_{q,\infty}^2 := 2\mathrm{Re}\Sigma_+(k,k)k = 2\hat{M}^2. \tag{3.5}$$

Even more, due to the fact that the squared dispersion relations of quarks and antiquarks are identical, this is also the asymptotic dispersion relation of antiquarks at negative ω.

The quality of the approximations is best estimated by a direct comparison of full and asymptotic dispersion relations, as done in Fig. 3.1. For $k > m_D$ or $2\hat{M}$ both dispersion relations are virtually indistinguishable. Since the main contributions to thermodynamic integrals are found at momenta k of order T, while m_D and M are of order gT (Eq. (2.19)), the asymptotic dispersion relations are good approximations of the full dispersion relations.

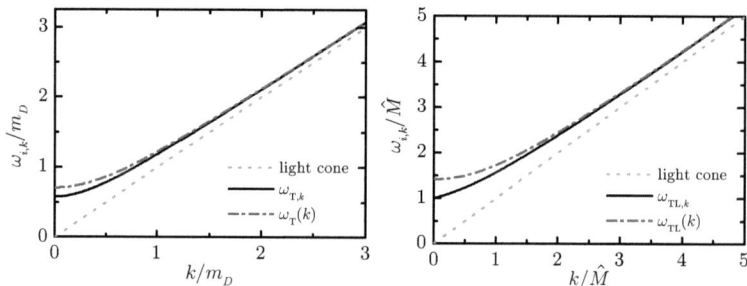

Figure 3.1: The full (solid lines) and asymptotic (dash dotted lines) dispersion relations for transverse gluons (left) and quarks (right) scaled by their respective mass parameters are shown as functions of the momentum k scaled by the same mass parameter.

These results are obtained and valid for the chiral limit. In order to compare with lattice calculations of quarks with nonzero quark restmasses they must be modified[1]. Assuming a current quark restmass $m_{q,\infty}$ of the order of the thermal quark mass $\sim gT$ or lower, this is done by introducing additional terms related to the restmass into the asymptotic mass or, equivalently, the dispersion relation [Pis89b, Pes98, Blu04]

$$\tilde{m}_{i,\infty}^2 = m_{i,0}^2 + \sqrt{2}m_{i,0}m_{i,\infty} + m_{i,\infty}^2. \tag{3.6}$$

The index i denotes either light quarks q with $m_{q,\infty}^2 = 2\hat{M}_q^2 = 2\hat{M}^2$, strange quarks s with $m_{s,\infty}^2 = 2\hat{M}_s^2 = 2\hat{M}^2|_{\mu=0}$ or gluons g, where $m_{g,0} = 0$ and therefore $\tilde{m}_{g,\infty}^2 = m_{g,\infty}^2 = m_D^2/2$. To keep our calculations consistent with the ones performed on the lattice [Che07, Baz09, DeT10][2], we also employ $m_{u,0} = m_{d,0} = m_{s,0}/10$ with $m_{s,0} = 105$ MeV [PDG08] which is roughly

---
[1]While it is also possible to include nonzero quark restmasses directly into the HTL propagators and self-energies, this causes gauge invariance to be lost and, additionally, complicates these expressions by introducing additional integrals [Kal84, Sei07, Sei09].

[2]That is, the concept dubbed "line of constant physics", referring to the almost physics quark masses.

compatible to the most recent estimates in [PDG10]. QPM results for the different values of the quark restmasses are virtually indistinguishable from each other. The overall effect of the quark restmasses depends on the quantity as well as temperature and chemical potential. For the interaction measure the maximum of $|(e-3p)_{m_s=105\text{ MeV}} - (e-3p)_{m_s=0}|/(e-3p)_{m_s=105\text{ MeV}}$ is 0.05.

A similar modification of the dispersion relation has been proposed [Sei07, Sei09], where the term $\sqrt{2}m_{i,0}m_{i,\infty}$ is omitted. It is stressed therein that this modification has a substantial impact on the extrapolation to the chiral limit. However, an evaluation of the change within the scope of this thesis has shown that it can be absorbed into a reparametrization of the effective coupling G^2 and has no significant effect for the extrapolation to large chemical potential considered here.

3.2 Net quark density

Focusing on the quasiparticle contributions we want to investigate the state variables in detail. We commence with the net quark density where the quasiparticle contribution is given from Eq. (2.53) as

$$n_{q,qp} = 2d_q \int_{d^4k} \left(\frac{\partial n_\text{F}}{\partial \mu} + \frac{\partial n_\text{F}^A}{\partial \mu} \right) \pi\Theta\left(\text{-Re}S_+^{-1}\right). \qquad (3.7)$$

We integrate by parts with respect to the energy integration, leading to $\delta(-\text{Re}S_+^{-1})$ as one factor[3], the single contributions of the particle species are extracted and realizing that $\partial n_\text{F}^{(A)}/\partial \mu = \underset{(+)}{(-)}\partial n_\text{F}^{(A)}/\partial \omega$ gives the other factor, for which we use that $n_\text{F} - n_\text{F}^A$ is symmetric in ω. The integrated part vanishes at the integration limits and the dispersion relations depend only on the absolute value of k so that

$$n_{q,qp} = \frac{d_q}{2\pi^2} \int_0^\infty dk\, k^2 \left[\left(n_\text{F} - n_\text{F}^A\right)\big|_{\omega_{\text{TL},k}} + \left(n_\text{F} - n_\text{F}^A\right)\big|_{\omega_{\text{Pl},k}} - \left(n_\text{F} - n_\text{F}^A\right)\big|_{\omega_{\text{Pl},k}^t} \right]. \qquad (3.8)$$

---

[3]While the Heaviside or step function is not differentiable at the jumps, neither in a classical nor weak sense, a derivative can be defined using the theory of distributions. In this sense, the derivative of the step function is the Dirac δ-distribution. Since all derivatives within the QPM are ultimately performed within thermodynamic integrals, we may use such δ-distributions as derivatives of step functions.

For an arbitrary function $f(\omega)$ with roots ω_k, we write in a "physicist's notation"

$$\begin{aligned}\frac{\partial \Theta(f)}{\partial \omega} &= \frac{\partial f}{\partial \omega} \frac{\partial \Theta(f)}{\partial f} \\ &= \frac{\partial f}{\partial \omega} \sum_k \frac{\delta(\omega - \omega_k)}{|-\partial f/\partial \omega|} \\ &= \sum_k \varepsilon\left(\partial f/\partial \omega\right) \delta(\omega - \omega_k),\end{aligned}$$

The combined distribution function $\sigma_n := (n_F - n_F^A)$ is monotonically decreasing for positive energies. As $\omega_{\text{Pl},k}^t < \omega_{\text{Pl},k}$, the contribution from the tachyonic plasmino branch is larger for all momenta, leading to an overall negative plasmino contribution to the net quark density. While the main contribution from the particles is fixed to thermal-like momenta, the momentum of the plasmino contributions scales approximately with $\sim \mu T$. The momentum of the main plasmino contributions thus increases along the characteristic curves from almost zero to about two times the momentum of the main particle contributions. Since the momenta are weighted with a factor of k^2, this eventually causes the negative plasmino contribution of the net quark density to cancel the positive particle contribution, leading to a negative net quark density.

For $\mu \gg T$ the contribution of the antiparticles vanishes from the integral and the particle distribution functions approach step functions at the Fermi surface, i.e. at the momenta $k_{i,\mu}$ where $\omega_{i,k_{i,\mu}} = \mu$.[4] Thus in the limit of $T \to 0$ we have

$$n_{q,qp} = \frac{d_q}{6\pi^2}\left[k_{\text{TL},\mu}^3 + k_{\text{Pl},\mu}^3 - (k_{\text{Pl},\mu}^t)^3\right], \qquad (3.9)$$

where the k_i can be interpreted as Fermi momenta.

Here, a simple argument can be given for the net plasmino density: since, at given momentum k, $\omega_{\text{Pl},k}^t < \omega_{\text{Pl},k}$ (cf. Fig. 2.6) and both dispersion relations are monotonically increasing with k (cf. Fig. 2.8) we find $k_{\text{Pl},\mu} < k_{\text{Pl},\mu}^t$ so that the overall contribution from the two plasmino

---

being well aware that in a strictly distributive sense we have for an infinitely differentiable test-function $t(\omega)$ with compact support (so that t vanishes at $\pm\infty$):

$$\int_{-\infty}^{+\infty} d\omega\, t \frac{\partial \Theta(f)}{\partial \omega} = -\int_{-\infty}^{+\infty} d\omega\, t'(\omega) \Theta(f(\omega))$$

$$= -\sum_{i \stackrel{\text{e.g.}}{=} [\omega_1,\omega_2]+..+[\omega_n,\infty)} \int_i d\omega\, t'(\omega)$$

$$\stackrel{\text{e.g.}}{=} -t(\omega_2) + t(\omega_1) - \ldots - t(\infty) + t(\omega_n)$$

$$= \int_{-\infty}^{+\infty} d\omega\, t \sum_k \varepsilon\,(\partial f/\partial \omega)\, \delta(\omega - \omega_k).$$

In the first step, the derivative was exchanged in the distributive sense, which amounts to integrating by parts with the integrated part vanishing due to t being on compact support. Assuming that $f(\omega)$ is continuous up to a finite number of even poles, its roots are alternatively occurring at the begin or end of integration intervals, i.e. at ω where $\Theta(f(\omega))$ either becomes 1 ($\varepsilon(\partial f/\partial \omega) = 1$) or 0 ($\varepsilon(\partial f/\partial \omega) = -1$). The above example represents one of the four cases $f(\omega \to \pm\infty) \lessgtr 0$ which lead to different interval configurations. All four cases can be comprised in the last expression, where each interval beginning/end is summed with the sign according to increasing/decreasing $f(\omega)$ with contributions at $\pm\infty$ vanishing due to t being on compact support. To make a connection with our previous result, the energy integral is finally reintroduced.

In the QPM, the functions f are linear functions of the real parts of the inverse propagators, which indeed are continuous up to the even plasmon and plasmino poles. The roots are the dispersion relations $\omega_k(T,\mu)$.

[4] In line with ω_k, the dependence of $k_{i,\mu}$ on μ is signified in the subscript rather than $k_i(\mu)$ as to expose that it is not a simple analytic function since it depends on the implicit functions ω_{i,k_i}.

3.3 Partial pressures

contributions turns out to be negative.

If assuming an asymptotic dispersion relation for the quark particle contribution, its net quark density reduces to the well-known expression

$$n_{q,qp,\text{TL}} = \frac{d_q}{6\pi^2} \left(\mu^2 - m_{q,\infty}^2\right)^{3/2} \tag{3.10}$$

which can be expanded for $m_{q,\infty} \ll \mu$ as

$$n_{q,qp,\text{TL}} = \frac{d_i \mu^3}{6\pi^2} \left(1 - 2\left(\frac{\alpha}{\pi}\right) + \frac{2}{3}\left(\frac{\alpha}{\pi}\right)^2 + ...\right), \tag{3.11}$$

where the asymptotic quark mass $m_{q,\infty}^2 = 2\tilde{C}_f G^2 = 4\mu^2 \alpha/(3\pi)$ (cf. Eq. (3.5) and Section 2.7) has been used.

3.3 Partial pressures

The investigation of the partial quasiparticle pressure is similar to the net quark density. Starting with the quark quasiparticle contributions from Eqs. (2.42)

$$p_{q,qp} = 2d_q \int_{d^3k} \int_0^\infty \frac{d\omega}{2\pi} \left(n_\text{F} + n_\text{F}^A\right) \left\{\pi\Theta(-\text{Re}S_+^{-1}) - \pi\Theta(+\text{Re}S_-^{-1})\right\} \tag{3.12}$$

we again integrate by parts with respect to ω and obtain three δ-distributions at the dispersion relations $\omega_{i,k}$. The integral of the combined distribution functions is given by $\sigma_p := -T \ln(n_\text{F} \exp(\omega/T)) - T \ln(n_\text{F}^A \exp(\omega/T))$ as can be verified by differentiation and has to be taken at the $\omega_{i,k}$. The integrated part vanishes at the integration limits, thus leaving us with

$$p_{q,qp} = \frac{d_q}{2\pi^2} \int_0^\infty dk\, k^2 \left[\sigma_p|_{\omega_{\text{TL},k}} + \sigma_p|_{\omega_{\text{Pl},k}} - \sigma_p|_{\omega_{\text{Pl},k}^t}\right]. \tag{3.13}$$

The function σ_p is strictly positive and monotonically decreasing so that, as for the net quark density, since $\omega_{\text{Pl},k}^t < \omega_{\text{Pl},k}$, the overall contribution of the plasminos turns out to be negative. It is also again due to the different scaling of the main contributions to the thermodynamic integral that the partial quasiparticle pressure of the quarks eventually turns negative along a characteristic curve.

The generalization of the above procedure to include the gluonic excitations is straightforward. As the change of the sign in the denominator of the statistical distribution functions does not affect the calculation we can introduce the short-hand notation

$$\begin{aligned} f_\pm &:= \frac{1}{e^\mp + S_i}, \\ e^\mp &:= e^{\beta(\omega_i \mp \mu_i)}, \end{aligned} \tag{3.14}$$

where the spin factor S_i is $+1$ for quarks and -1 for gluons. The dependence of $f_\pm$ and $e^\mp$ on the quasiparticle species $i = g, q$ (and also s) is implied. In the limit $\mu = 0$, $e^+ = e^-$ is abbreviated as e and $f_+ = f_-$ as f. Rewriting $\sigma_p = -S_i T[\ln(f_+ e) + \ln(f_- e)]$ in terms of $f_\pm$ and $e^\mp$, which contains a factor 2 for the gluons, then gives the general expression for the pure quasiparticle partial pressure

$$p_{qp} = \frac{1}{2\pi^2} \int_0^\infty dk\, k^2 \Bigg[d_q\, \sigma_p\big|_{\omega_{\text{TL},k}} + d_q\, \sigma_p\big|_{\omega_{\text{Pl},k}} - d_q\, \sigma_p\big|_{\omega^t_{\text{Pl},k}} \\ + d_g\, \sigma_p\big|_{\omega_{\text{T},k}} + \frac{d_g}{2} \sigma_p\big|_{\omega_{\text{L},k}} - \frac{d_g}{2} \sigma_p\big|_{\omega^t_{\text{L},k}} \Bigg]. \quad (3.15)$$

With σ_p keeping its properties in the gluonic case[5], it is clear that the tachyonic longitudinal mode also gives a negative contribution larger than the one from the regular longitudinal mode resulting in an overall negative contribution from the longitudinal mode to the quasiparticle pressure. The transverse gluon contribution, however, turns out be larger for all (T, μ) leading to a positive gluon quasiparticle pressure.

Concentrating on the quark particle and transverse gluon quasiparticle contributions for the following approximations, we assume asymptotic dispersion relations $\omega_i^2(k) = k^2 + m_{i,\infty}^2$. Integrating Eq. (3.15) by parts with respect to the momentum k then gives

$$p_{i,qp} = \frac{d_i}{6\pi^2} \int_0^\infty dk \frac{k^4}{\omega_i(k)} \big[f_+ + f_-\big]. \quad (3.16)$$

In order to check the perturbative limit of the quasiparticle model we derive the zero temperature expression for the particle contribution to the quark quasiparticle partial pressure. At $T = 0$, the antiparticle contribution from $f_- = n_{\bar{\text{F}}}^A$ vanishes and $f_+ = n_{\text{F}} \to \Theta(\mu - \omega_{\text{TL}}(k)) = \Theta(\sqrt{\mu^2 - m_\infty^2} - k)$, i.e. the integral is cut off at the Fermi momentum $k_{\text{F}} = \sqrt{\mu^2 - m_\infty^2}$. Integration by substitution, using $k = m_\infty \sinh x$, then gives

$$p_{q,qp,\text{TL}} = \frac{d_q}{6\pi^2} \frac{m_{q,\infty}^4}{8} \left\{ 2\left(\frac{\mu^2}{m_{q,\infty}^2} - 1\right)^{\frac{3}{2}} \frac{\mu}{m_{q,\infty}} - 3\left(\frac{\mu^2}{m_{q,\infty}^2} - 1\right)^{\frac{1}{2}} \frac{\mu}{m_{q,\infty}} + 3\ln\left(\sqrt{\frac{\mu^2}{m_{q,\infty}^2} - 1} + \frac{\mu}{m_{q,\infty}}\right) \right\} \quad (3.17)$$

which can be approximated for $m_{q,\infty} \ll \mu$ by

$$p_{q,qp,\text{TL}} = \frac{d_q}{48\pi^2} \left\{ 2\mu^4 - 6\mu^2 m_{q,\infty}^2 + \frac{9}{4} m_{q,\infty}^4 + 3 m_{q,\infty}^4 \ln \frac{2\mu}{m_{q,\infty}} \right\}, \quad (3.18)$$

where, for $\mu \sim 0.8$ GeV, the logarithmic term contributes about one quarter for up and down quarks and up to two thirds for strange quarks. Assuming the asymptotic quark mass

---
[5]Note however, that the domain of σ_p for gluons is $[0, \infty)$ as opposed to the quarks.

3.4 Mean field pressure contribution

$m_{q,\infty}^2 = 4\mu^2 \alpha/(3\pi)$ then gives

$$p_{q,qp,\text{TL}}(\mu) = \frac{N_f \mu^4}{4\pi^2}\left(1 - 4\left(\frac{\alpha}{\pi}\right) + \left(2 + \frac{4}{3}\ln 3 - \frac{4}{3}\ln\frac{\alpha}{\pi}\right)\left(\frac{\alpha}{\pi}\right)^2\right). \qquad (3.19)$$

For the transverse gluons $\sigma_p = 2T\ln(fe) = -2T\ln(1 - e^{-\omega_T(k)/T})$ which, at small temperatures $T \ll \omega_T(k)$, can be approximated by $2Te^{-\omega_T(k)/T}$. As can be expected, all gluon quasiparticle partial pressures vanish in the limit $T \to 0$: by again assuming an asymptotic dispersion relation and substituting $k = m_{g,\infty}\sinh x$, we obtain

$$p_{g,qp,\text{T}} = \frac{d_g}{2\pi^2}m_{g,\infty}^2 T^2 K_2\left(\frac{m_{g,\infty}}{T}\right) \xrightarrow{T \ll m_{i,\infty}} \frac{d_g}{2\pi^2}m_{g,\infty}^2 T^2 \sqrt{\frac{\pi T}{2m_{g,\infty}}}e^{-\frac{m_{g,\infty}}{T}}, \qquad (3.20)$$

where K_2 is the second modified Bessel function. The expression the partial quasiparticle pressure contribution of the transverse gluons vanishes for $T \to 0$. This also holds for the overall gluon partial pressure (2.42) since $n_B(T \to 0) = 0$.

3.4 Mean field pressure contribution

The mean field pressure B follows from Eqs. (2.61), (2.52) and (2.45) as

$$B = B_c + \sum \underbrace{\int_{T_c}^T dT' \frac{\partial p_i}{\partial \Pi_i}\frac{\partial \Pi_i}{\partial T'} + \int_0^\mu d\mu' \frac{\partial p_i}{\partial \Pi_i}\frac{\partial \Pi_i}{\partial T'}}_{B_i(T)}. \qquad (3.21)$$

For the quasiparticle contribution of transverse gluons and quark particles with asymptotic dispersion relations, the implicit dependence of the partial pressures on the temperature is via the asymptotic mass:

$$B_{i,qp} = \int_{T_c}^T dT' \frac{\partial p_{i,qp}}{\partial m_{i,\infty}^2}\frac{\partial m_{i,\infty}^2}{\partial T'} + \int_0^\mu d\mu' \frac{\partial p_{i,qp}}{\partial m_{i,\infty}^2}\frac{\partial m_{i,\infty}^2}{\partial \mu'}. \qquad (3.22)$$

Using Eq. (3.15) for the quasiparticle contributions to the partial pressure the derivative with respect to the asymptotic mass acts on σ_p giving $[f_+ + f_-]/2\omega_i$ so that

$$\frac{\partial p_{i,qp}}{\partial m_{i,\infty}^2} = -\frac{d_i}{4\pi^2}\int dk \frac{k^2}{\omega_i(k)}\left[f_+ + f_-\right]. \qquad (3.23)$$

At vanishing chemical potential the squared bracket [] simplifies to $2f$ giving. e.g. for the transverse gluon case, the well-known expression [Pes96]

$$B_{g,qp,\text{T}}(T) = -\frac{d_g}{2\pi^2}\int_{T_c}^T dT' \frac{\partial m_{g,\infty}^2}{\partial T'}\int_0^\infty dk k^2 \frac{n_B(\omega_T(k))}{\omega_T(k)}. \qquad (3.24)$$

The latter may also be derived directly from the HTL expression (3.21) and the HTL partial pressures (2.42), where the derivative with respect to the real part of the self-energies yield δ-distributions at the dispersion relations. Letting $\mathrm{Re}\Pi_\mathrm{T}|_{\omega_{\mathrm{T},k}} = \omega_{\mathrm{T},k}^2 - k^2 = m_{g,\infty}^2 = m_D^2/2$ yields the result.

At vanishing temperature the quasiparticle contribution of the gluons vanishes due to $n_\mathrm{B}(T \to 0) = 0$. The contribution from the quark particles to the mean field pressure can be approximated in the same way as the quasiparticle net density and the quasiparticle partial pressure by realizing the distributions functions to be step functions and substituting $k = m_{i,\infty} \sinh x$. The result for the derivative of the quasiparticle partial pressure with respect to the asymptotic mass is

$$\frac{\partial p_{q,\mathrm{qp,TL}}}{\partial m_{i,\infty}^2} = -\frac{d_i}{8\pi^2}\left[\mu^2 \sqrt{\mu^2 - m_{q,\infty}^2} - m_{q,\infty}^2 \ln\left(\sqrt{\frac{\mu^2}{m_{q,\infty}^2} - 1} + \frac{\mu}{m_{q,\infty}}\right)\right]. \quad (3.25)$$

Additionally assuming small asymptotic masses $m_{q,\infty}^2 = 4\mu^2\alpha/(3\pi) \ll \mu$ and neglecting the μ-dependence of the coupling in order to analytically solve the integral with respect to the chemical potential yields

$$B_{q,qp,\mathrm{TL}}(\mu) = C - \frac{N_f \mu^4}{4\pi^2}\left(2\left(\frac{\alpha}{\pi}\right) - \left(\frac{\alpha}{\pi}\right)^2 \left(\frac{4}{3} + \frac{4}{3}\ln 3 - \frac{4}{3}\ln\frac{\alpha}{\pi}\right)\right). \quad (3.26)$$

Thus, for a plasma described by N_f quasiquark flavors we find from Eqs. (3.19) and (3.26) the following perturbative expression for the pressure at vanishing temperature:

$$p_{qp}(T=0) = \frac{N_f \mu^4}{4\pi^2}\left\{1 - 2\left(\frac{\alpha}{\pi}\right) + \frac{2}{3}\left(\frac{\alpha}{\pi}\right)^2\right\} - C. \quad (3.27)$$

This is, of course, the same result which follows from integrating the expansion of the quasiparticle contribution to the net quark density (3.11) with respect to the chemical potential.

The coefficient of the $\mathcal{O}(\alpha)$ term equals the strictly perturbative results in [Fre77, Fra01, Fra02]. The coefficient of the next-order term deviates from the perturbation expansion (prefactor 12.1), providing a middle way between the latter and the hard-dense-loop approach in [AS02] (prefactor 0.06).

3.5 Entropy and energy density

As for the other thermodynamic quantities, the quasiparticle contributions to the entropy density are most easily extracted from Eqs. (2.40) and (2.41) via an integration by parts with respect to the energy density. While the Θ-function is differentiated to δ-distributions, the

3.5 Entropy and energy density

statistical distribution functions are integrated to a dimensionless function

$$\sigma_s := \underbrace{\beta\omega(f_+ + f_-)}_{=:\sigma_e/T} \underbrace{-\beta\mu(f_+ - f_-)}_{-\mu\sigma_n/T} \underbrace{-S_i(\ln(f_+e) + \ln(f_-e))}_{\sigma_p/T} \quad (3.28)$$

again containing a factor 2 for the gluons. The ω-integration then breaks down at the δ-distributions and gives σ_s at the dispersion relations with signs according to the slope of $\mathrm{Re}\Pi_i(\omega)$. The resulting quasiparticle entropy density expression is

$$\begin{aligned} s_{qp} = \frac{1}{2\pi^2} \int_0^\infty \mathrm{d}k\, k^2 \Big[& d_q\, \sigma_s|_{\omega_{\mathrm{TL},k}} + d_q\, \sigma_s|_{\omega_{\mathrm{Pl},k}} - d_q\, \sigma_s|_{\omega_{\mathrm{Pl},k}^t} \\ & + d_g\, \sigma_s|_{\omega_{\mathrm{T},k}} + \frac{d_g}{2}\sigma_s|_{\omega_{\mathrm{L},k}} - \frac{d_g}{2}\sigma_s|_{\omega_{\mathrm{L},k}^t} \Big]. \end{aligned} \quad (3.29)$$

Once more, the integrated variable σ_s is a strictly positive function which in the domain $[0, \infty)$ is monotonically decreasing with asymptote 0. Thus, the argument of negative contributions from plasminos and longitudinal gluons also holds for the quasiparticle entropy density. The overall quasiparticle entropy density does, however, stay positive along the characteristics.

The expression simplifies if assuming an asymptotic dispersion relation. Considering the case of transverse gluons and quark particles with asymptotic masses $m_{i,\infty}^2$ and integrating by parts the logarithmic terms in Eq. (3.29) with Eq. (3.28), corresponding and analogously to the pressure, gives the expression

$$s_{i,qp} = \frac{d_i}{2\pi} \int_0^\infty \mathrm{d}k\, k^2 \left\{ \frac{\frac{4}{3}k^2 + m_{i,\infty}^2}{\omega_i(k)T}\left[f_+ + f_-\right] - \frac{\mu_i}{T}\left[f_+ - f_-\right] \right\}. \quad (3.30)$$

The quasiparticle contribution to the energy density follows directly from the interpretation of Eq. (3.28), where $\sigma_e = \omega(f_+ + f_-)$, and Eqs. (3.7), (3.15) and (3.29) as

$$\begin{aligned} e_{qp} = \frac{1}{2\pi^2} \int_0^\infty \mathrm{d}k\, k^2 \Big[& d_q\, \sigma_e|_{\omega_{\mathrm{TL},k}} + d_q\, \sigma_e|_{\omega_{\mathrm{Pl},k}} - d_q\, \sigma_e|_{\omega_{\mathrm{Pl},k}^t} \\ & + d_g\, \sigma_e|_{\omega_{\mathrm{T},k}} + \frac{d_g}{2}\sigma_e|_{\omega_{\mathrm{L},k}} - \frac{d_g}{2}\sigma_e|_{\omega_{\mathrm{L},k}^t} \Big], \end{aligned} \quad (3.31)$$

as expected from general thermodynamics. Since the antiparticle contribution vanishes with $T \to 0$, an expansion closely related to the one for the pressure can be performed.

In the limit $T \to 0$ the energy density of each quasiparticle species follows from general thermodynamics as $e_i = -p_i + \mu_i n_i$, since – even for finite entropy density – the contribution sT vanishes. However, using this relation it is clear that the entropy density itself, due to its connection to the other three quantities via Eq. (3.28), has to vanish at $T = 0$. Thus, Nernst's theorem is obeyed by the quasiparticle entropy density. This also holds for the full HTL QPM entropy.

4 The effective quasiparticle model

In the preceding chapters the HTL QPM was derived and investigated analytically. This was done in order to provide a systematic series of approximations and assumptions leading from actual QCD to the established quasiparticle model, as e.g. summarized in [BKS07a], and – in the reverse direction – to provide a coherent path for the improvement of the latter. In this chapter we complete this connection by specifying the necessary assumptions. The established model is then adjusted to the most recent lattice results and the properties of the extrapolated quantities are investigated as well as compared to lattice predictions for the extrapolation (for instance, the Taylor coefficients for the pressure and the chiral phase transition curve). On the general side, the self-consistency of the model is verified and some aspects of quasiparticle models are demonstrated.

4.1 Necessary approximations

In many cases it proves useful to use a simplified and analytically more accessible version of the HTL quasiparticle model which we dub effective quasiparticle model (eQPM). This model is well established and has been presented e.g. in [Blu04, BKS07a] as standard Rossendorf QPM[1]. It therefore provides the reference point for the improvements of the HTL QPM.

The eQPM follows from the HTL QPM given the following assumptions:

1. the collective excitations, plasmons and (anti)plasminos, are exponentially suppressed[2] and can therefore be neglected,

2. the quasiparticle widths as well as damping effects (i.e. the imaginary parts of the self-energies) are small and can be ignored,

3. the full dispersion relations, only given implicitly (cf. Section 2.5), can be approximated by analytic expressions $\tilde{\omega}_i^2 = k^2 + \tilde{m}_{i,\infty}^2$ with asymptotic masses $\tilde{m}_{i,\infty} = \tilde{m}_{i,\infty}(T,\mu)$ for

---

[1]The original quasiparticle model as published in [Pes94, Pes96] (for gluons only) and [Pes00] (including quarks) is found from this model by assuming vanishing quark contributions or quark restmasses, respectively.

[2]That means, after computing the propagators using Dyson's relation from the HTL self-energies, the residues of the poles in the spectral density of both plasmon and (anti)plasmino propagators vanish exponentially for momenta $k \sim T, \mu$ which give the dominant main contributions to thermodynamic integrals.

thermal-like momenta as relevant in thermodynamic integrals.

The negligence of collective modes and damping effects is suggested by the fact that both are bound to the medium frame of reference and thus show only minimal effects on particles at high momenta [Pes98]. This is verified in Chapter 5.

Taking all three assumptions into account, the eQPM expressions for the thermodynamic quantities equal the Eqs. (3.16), (3.22), (3.23), (3.30) and (3.8) for the pure quasiparticle parts in Chapter 3 with the modified asymptotic mass $\tilde{m}_{i,\infty}^2$, i.e. modified dispersion relation $\tilde{\omega}_i^2 = k^2 + \tilde{m}_{i,\infty}^2$:

$$\begin{aligned}
p_i^{eQP} &= \frac{d_i}{6\pi^2} \int_0^\infty dk \frac{k^4}{\tilde{\omega}_i(k)} \left[f_+ + f_- \right], \\
B_i^{eQP} &= \int_{T_c}^T dT' \frac{\partial p_i^{eQP}}{\partial \tilde{m}_{i,\infty}^2} \frac{\partial \tilde{m}_{i,\infty}^2}{\partial T'} + \int_0^\mu d\mu' \frac{\partial p_i^{eQP}}{\partial \tilde{m}_{i,\infty}^2} \frac{\partial \tilde{m}_{i,\infty}^2}{\partial \mu'}, \\
\frac{\partial p_i^{eQP}}{\partial \tilde{m}_{i,\infty}^2} &= -\frac{d_i}{4\pi^2} \int dk \frac{k^2}{\tilde{\omega}_i(k)} \left[f_+ + f_- \right], \\
s_i^{eQP} &= \frac{d_i}{2\pi} \int_0^\infty dk\, k^2 \left\{ \frac{\frac{4}{3}k^2 + \tilde{m}_{i,\infty}^2}{\tilde{\omega}_i(k)T} \left[f_+ + f_- \right] - \frac{\mu_i}{T} \left[f_+ - f_- \right] \right\}, \\
n_q^{eQP} &= \frac{d_q}{2\pi^2} \int_0^\infty dk\, k^2 \left[f_+ - f_- \right]. \quad (4.1)
\end{aligned}$$

As for the HTL QPM, the overall pressure then follows from $p^{eQP} = \sum p_i^{eQP} - B^{eQP}$ (cf. Eq. (2.44)) while the overall entropy density and energy density follow by simple summation.

In comparison with the HTL QPM the eQPM is much simpler as its partial contributions look like simple ideal gas expressions with modified masses so that only one phase space integral has to be solved. In contrast, the HTL QPM requires a two-dimensional integration with respect to energy and momentum – demanding a good deal more computing time – and, due to the damping terms, is much more difficult to handle analytically. Simple limit discussions are almost always only possible for the pure quasiparticle contributions to the HTL expressions – which for asymptotic dispersion relations lead to the eQPM expressions (cf. Chapter 3).

The introduction of the constant restmass via the modified dispersion relation causes no differences for the arguments put forward in Chapter 3, so that with $\omega(k) \to \tilde{\omega}(k)$ the results are valid for the eQPM. Also, as the implicit dependence of the self-energies on temperature and chemical potential is handed down to the thermal quasiparticle masses, the expressions for the flow equation and the mean field pressure along a characteristic curve in Sections 2.10 and 2.11 can be employed with the simple change $\Pi_i \to \tilde{m}_{i,\infty}^2$.

4.2 Comparison with lattice results

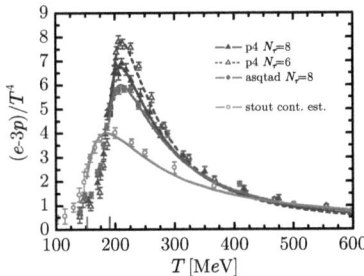

Figure 4.1: Comparison of the eQPM interaction measure to the lattice QCD data (symbols) for lattice actions p4 (blue), asqtad (red) and stout (green) and lattice spacings N_τ from [Che07, Baz09, Bor10b]. Solid (dashed) curves are for $N_\tau = 8$ (6). The statistical errors from the lattice are indicated.

4.2 Comparison with lattice results

As outlined in Section 2.12, the model parameters, i.e. λ and T_s for the effective coupling and the pressure integration constant B_c, have to be fixed by comparing with lattice results. We use the most recent results of two major lattice collaborations: the hotQCD collaboration [Che07, Baz09] and the Wuppertal-Budapest collaboration [Bor10a, Bor10b].

We refer to the results by the lattice action and temporal lattice extent N_τ of the particular calculations. Evaluations with larger N_τ are closer to the continuum limit and are therefore preferred. While the hotQCD collaboration uses the p4 and asqtad actions with $N_\tau = 6$ and 8 [Che07, Baz09], the Wuppertal-Budapest collaboration relies on the stout action with N_τ up to 12 [Bor10a, Bor10b]. In addition, the latter group therein also provides a continuum estimate which we use here.

The parameter adjustment is performed at vanishing chemical potential by χ^2-minimization of the difference of the scaled eQPM interaction measure $\Delta^{eQP}/T^4 = (e^{eQP} - 3p^{eQP})/T^4 = s^{eQP}/T^3 - 4p^{eQP}/T^4$ to the lattice results, where the pressure constant is fixed for any combination of (λ, T_s) by requiring the pressure p^{eQP} at T_c to be equal to the lattice pressure $p_{\text{lattice}}(T_c)$ (cf. Section 2.12). The results are shown in Fig. 4.1 and the corresponding model parameters are given in Tab. 4.1. Despite the obvious differences between the different lattice results, which are attributed to different discretization artifacts, the model is flexible enough to allow for an accurate description of all four. One may assume the quality of the lattice description to increase from the p4 lattice action to the asqtad and then to the stout action as well as to grow with rising N_τ. The seemingly large χ^2 for the asqtad action is due to the lack of results above $T \sim 450$ MeV.

In the left and right panels of Fig. 4.2 we compare the thermodynamic state variables of the eQPM with the ones obtained on the lattice for the stout action and the p4 action with $N_\tau = 8$. The eQPM results match the results from the lattice very closely, with the larger deviations visible for the case of the p4 action with $N_\tau = 8$. The differences are due to the variation in the

action	N_τ	$p(T_c)/T_c^4$	T_s [MeV]	λ [MeV]	B_c	χ^2/dof
p4	6	0.58	173	15	$-(145\text{ MeV})^4$	3.09
p4	8	0.70	154	24	$(162\text{ MeV})^4$	1.21
asqtad	8	0.76	139	35	$(158\text{ MeV})^4$	3.35
stout	∞ (est.)	0.63	48	83	$(75\text{ MeV})^4$	0.95

Table 4.1: Parameters of the eQPM as a result of adjusting the eQPM interaction measure to the lattice QCD results [Che07, Baz09, Bor10b] with the eQPM pressure $p(T_c)$ being fixed to the lattice pressure p_lattice at T_c (cf. Section 2.12).

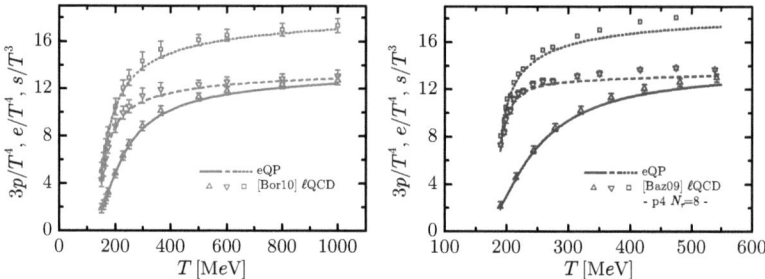

Figure 4.2: Comparison of scaled pressure $3p/T^4$ (upright triangles, solid lines), scaled energy density e/T^4 (reversed triangles, dashed lines) and scaled entropy density s/T^3 (circles, dotted lines) derived from the interaction measure on the lattice (symbols) and via the eQPM (curves). Left panel: Lattice results from [Bor10b]. Right Panel: Lattice results for the p4 action with $N_\tau = 8$ from [Baz09]. Statistical errors from the lattice are indicated where available.

peak position and height in Δ/T^4 for the actions as well as the fact that the eQPM interaction measure does not replicate the lattice results with absolute perfection.

From Eqs. (2.67) is clear, that more extensive peaks lead to larger pressures as well as that the differences in the pressure enter the entropy density and energy density with factors of 4 and 3, respectively. Therefore, while the deviation of the pressure is only small, some larger variation is noticeable in the entropy density. One possible explanation for the better description of the lattice result from [Bor10b] by the eQPM (as a thermodynamic model) in comparison to the one from [Baz09] may be, that the thermodynamics of the lattice calculations continue to improve with better lattice actions and increasing lattice extents.

In general, the region of large temperatures is well understood perturbatively and, via the 1-loop self-energies in the HTL approximation, also provides the basis for our model. Thus, the compatibility of eQPM and lattice results in the high-temperature regime is to be expected. In contrast, it is very surprising that the introduction of just one additional parameter T_s suffices to allow a description of the nonperturbative region close to the phase transition. The small number of parameters is one of the great advantages of the eQPM when compared to e.g. (P)NJL models with a multitude of parameters, and may be considered as a hint that the quasiparticle degrees of freedom present a good approximation.

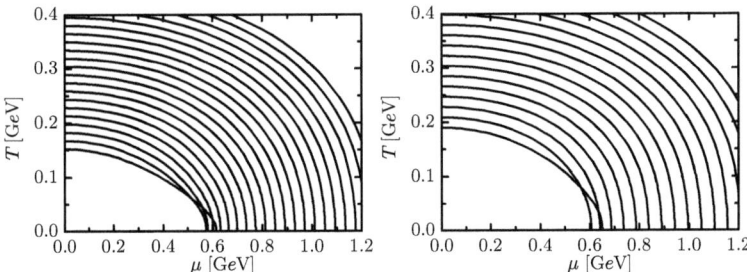

Figure 4.3: Some characteristic curves $(T(x), \mu(x))$ of the flow equation (2.56), emerging from the temperature interval $T_0 \epsilon [T_c, 450\,\text{MeV}]$, are shown in the T-μ-plane. The left curves are for [Bor10b] lattice results with $T_c = 152$ MeV, the right ones for [Baz09] results using the p4 action with $N_\tau = 8$ and $T_c = 190$ MeV.

4.3 Extrapolation to nonzero chemical potential

Fixing the model parameters at $\mu = 0$ to match the lattice results provides the boundary condition necessary to solve the Cauchy problem presented by the system of ordinary differential equations (2.58) that follow from the flow equation (2.56) by applying the method of characteristics (cf. Section 2.10). The coefficients of the eQPM flow equation are given in Appendix D.

Fig. 4.3 shows the characteristics of the eQPM using two of the adjustments to lattice results from Section 4.2. Since the the eQPM describes quark and gluon degrees of freedom, the model is limited to the temperature interval $[T_c, \infty)$ at vanishing chemical potential. Consequently, only those regions of the phase diagram, which are populated by characteristic curves $(T(x), \mu(x))$ emerging from this interval, are accessible to the eQPM.

As a consequence, for the [Bor10b] lattice results with $T_c = 152$ MeV this region turns out larger than the region for the case of [Baz09] lattice results for the p4 action with $N_\tau = 8$ with $T_c = 190$ MeV. The minimum accessible chemical potentials are 576 MeV and 606 MeV, respectively, revealing a difference comparable to the discrepancy in T_c.

The characteristics of the eQPM are known to exhibit crossings for adjustments to older lattice results [Pes00, Pes02]. For the adjustments to recent lattice results regions of crossing characteristics appear at small temperatures close to the pseudocritical characteristic as well. It is interesting to note that the region of the crossings, with extent of about 50 MeV in T- and μ-direction, also moves in the direction of the chemical potential with approximately the difference in T_c. A rough parametrization of the region independent of the chosen lattice adjustment can be given by $\mu = 600 \pm 50$ MeV and $T \epsilon [0, 50 \pm 20]$ MeV. As outlined in Section 2.10, the maximum set of consistent characteristics is chosen for the description of the this region, thus ensuring conservation laws.

Figures 4.4 through 4.10 show the thermodynamic quantities as results of the extrapolation

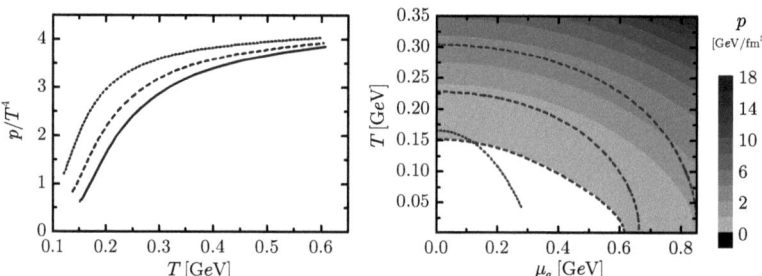

Figure 4.4: Left panel: The scaled pressure p/T^4 of the eQPM adjusted to [Bor10b] lattice results is shown as a function of the temperature T for chemical potential $\mu = 0$ (solid curve), 0.2 GeV (dashed curve) and 0.3 GeV (dotted curve). Right panel: Contour plot of the pressure p. The characteristic curves emerging from $T_0/T_c = (1.0, 1.5, 2.0)$ are shown as dashed curves. For reference, the predicted chemical freeze-out line from the statistical model in [Cle06] is plotted (dotted curve).

procedure using the lattice results from [Bor10b] as boundary values within the T-μ-plane as functions of the temperature at several values of the chemical potential and additionally as contour plots. All considered state variables of the eQPM, pressure, interaction measure, entropy density, net quark density and energy density, increase with rising chemical potential at constant temperature. The mean field pressure B increases with μ at smaller temperatures while it starts to decrease if considering larger temperatures. The effective coupling G^2 decreases with increasing chemical potential at constant temperature.

The contour plots show that no irregularities arise from the extrapolation procedure and that, indeed, the information obtained at vanishing chemical potential is transported in a thermodynamically consistent way to finite μ and even to vanishing temperature. Thus, the thermodynamic quantities at $T = 0$, as functions of μ, are similar to the ones as functions of T at $\mu = 0$ with specific scalings and shifts. For example, the contour plots also reveal that some quantities, such as the interaction measure $e - 3p$ (Fig. 4.5) and the pure mean field pressure $B - B_0$ (Fig. 4.9), stay almost constant along the path of the characteristic curves, especially for characteristics emerging from $T_0 > 1.5T_c$. In a special case, this can be shown analytically for the latter (cf. Section 4.7). It is also worth noting that the lines of constant pressure and the lines of constant energy density closely resemble each other (cf. Fig. 4.4 and 4.8). The limits $n_q(\mu \to 0) = 0$ (Fig. 4.7) and $s(T \to 0) = 0$ (Fig. 4.6) are visible.

4.4 Check of model consistency

Along each characteristic, and thus within the whole accessible part of the T-μ-plane, the flow equation provides the effective coupling in a way that ensures thermodynamic consistency, i.e. the overall pressure $p = \sum p_i - B$ represents a genuine potential via the mean field pressure

4.4 Check of model consistency

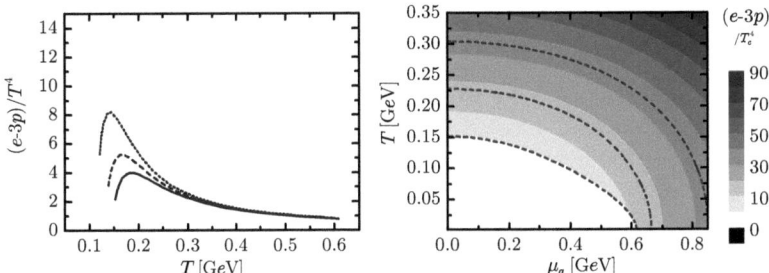

Figure 4.5: As Fig. 4.4 but for the scaled interaction measure $(e-3p)/T^4$.

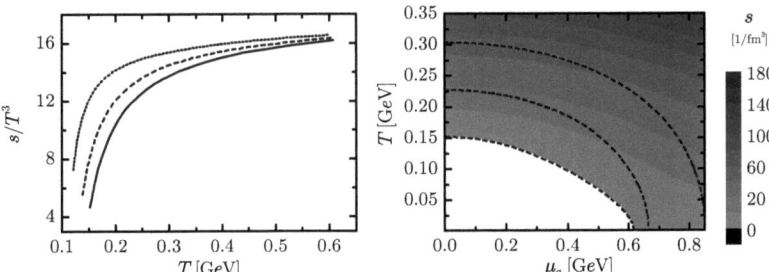

Figure 4.6: As Fig. 4.4 but for the scaled entropy density s/T^3.

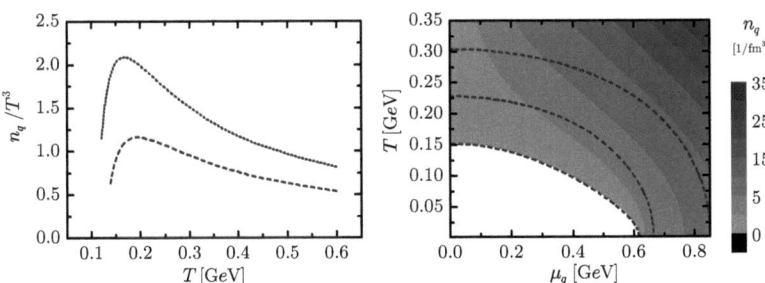

Figure 4.7: As Fig. 4.4 but for the scaled net quark density n_q/T^3.

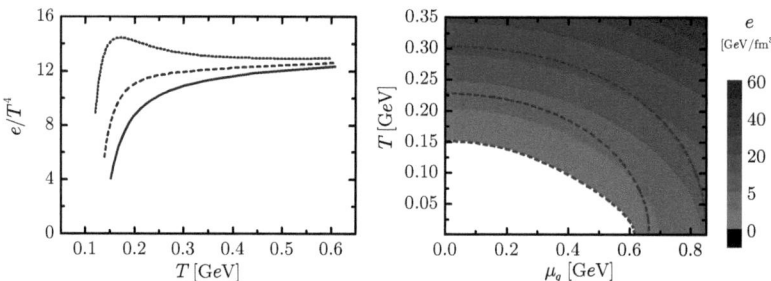

Figure 4.8: As Fig. 4.4 but for the scaled energy density e/T^4. Note that the peak structure appearing for larger values of the chemical potential are due to the scaling with T^4.

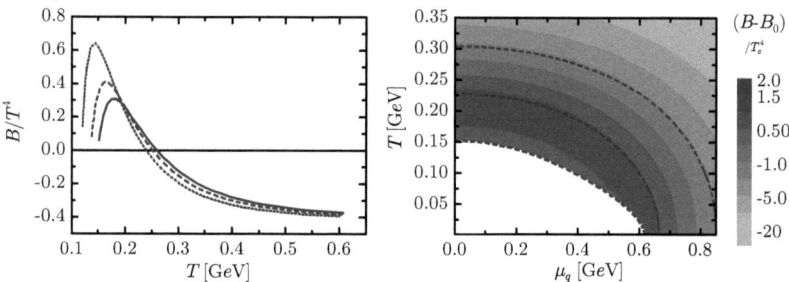

Figure 4.9: As Fig. 4.4 but for the scaled mean field pressure B/T^4.

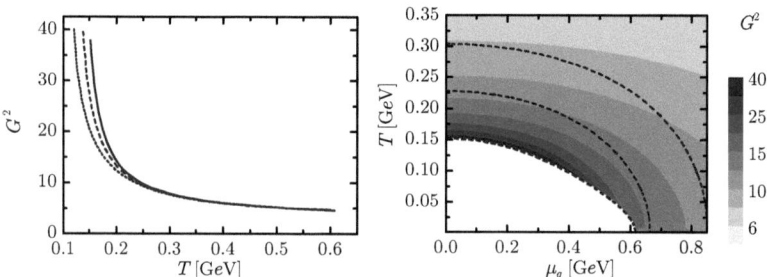

Figure 4.10: As Fig. 4.4 but for the effective coupling G^2.

4.4 Check of model consistency

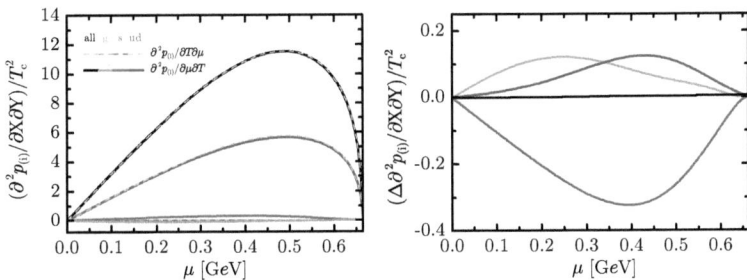

Figure 4.11: The absolute values (left panel) and difference of (right panel) the second derivatives of the pressure (parts) $p_{(i)}$ with respect to temperature T and chemical potential μ are shown along one characteristic curve emerging from $T = 1.5 T_c$. The calculation was performed for the parameters matching the lattice results from [Bor10b] where $T_c = 152$ MeV. The integrability condition (2.48) of the pressure is satisfied by the overall pressure p (black solid and grey dotted curves) only. As predicted, noticeable deviations occur for the partial contributions $p_{(i)}$ (cf. Eq. (2.51)) of gluons (yellow curves), up/down quarks (red solid and grey dashed curves) and – most prominent – strange quarks (blue curves). They add up to zero for the overall pressure.

B (cf. Section 2.9). This can be checked by numerically evaluating Eq. (2.48) between (at least) two very close characteristics. Fig. 4.11 shows the absolute values and the difference of the second derivatives with respect to temperature and chemical potential of the pressure p and its parts $p_{(i)}$ using the parametrization for [Bor10b] lattice results. As can be seen, the overall pressure satisfies the integrability condition and therefore constitutes a thermodynamic potential.

As outlined in Section 2.9, the same is not true for the partial contributions $p_{(i)} := p_i - B_i$ (cf. Eq. (2.51)); the reason being that the integrability condition is ensured by the flow equation for the overall pressure $p = \sum p_{(i)}$ (and overall mean field pressure $B = \sum B_i = -\sum p_{(i)} - p_i$) only and consequently the partial contributions $p_{(i)}$ to the overall pressure do not represent thermodynamic potentials. On the other hand it is essential to investigate the single contributions to a thermodynamic quantity such as the energy density which are derived from exactly these $p_{(i)}$ and B_i. It is therefore useful to investigate the deviations occurring in the $p_{(i)}$ due to the global nature of the flow equation.

The differences in the second derivatives with respect to T and μ are shown in Fig. 4.11. For the light quarks we find rather small relative deviations of about 2.5 percent at maximum for the parametrization describing [Bor10b] lattice results and about 3.5 percent for the parametrization describing p4 $N_\tau = 8$ lattice results from [Baz09]. While the absolute deviations of the second derivatives of gluons and strange quarks are quite small, too, the relative deviations of these two contributions can be sizeable since one of the second derivatives is manifestly zero while the other is, in general, nonzero[3]. Investigating the relative differences for the parametrization to

---
[3] From $n_g = n_s = 0$ we have $\partial^2 p_{(g/s)}/\partial T \partial \mu = \partial n_{g/s}/\partial T = 0$ while $\partial^2 p_{(g/s)}/\partial \mu \partial T = \partial s_{g/s}/\partial \mu \neq 0$ due to implicit dependencies of the partial entropy densities on the chemical potential.

[Baz09] lattice results for the p4 action with $N_\tau = 8$ yields similar results with smaller strange quark, enlarged gluon and negative up/down quark contribution to $\Delta \partial^2 p_{(i)}/\partial X \partial Y$, i.e. the qualitative behavior of the deviations is not fixed and depends on the parametrization.

One persistent feature is that for vanishing chemical potential, where the model has to be self-consistent per definitionem, and for vanishing temperature the deviation between the second derivatives of the partial contributions goes to zero. The latter is, on the one hand, due to vanishing partial entropy densities $s_i = \partial p_{(i)}/\partial T$ and thus first derivatives of the $p_{(i)}$ at $T = 0$ (Nernst's law, cf. Section 3.5) while, on the other hand, a similar argument as for Nernst's law can also be made for the first derivative of the net particle densities with respect to the temperature, i.e. $\partial n_i/\partial T = \partial^2 p_{(i)}/\partial T \partial \mu = 0$ at $T = 0$.

The partial net particle and partial entropy densities can directly be determined via analytic expressions from the effective coupling, however, for $\mu > 0$ and $T > 0$ do not satisfy a Maxwell relation $\partial s_i/\partial \mu = \partial n_i/\partial T$. Consequently, the pressure parts $p_{(i)}$ as integrals of the s_i/n_i (similar to Eq. (2.65)) are path-dependent and so are any other partial quantities derived from the pressure. While the deviations are rather small for the single quantities, they can, depending on the paths considered, become large in the integrated quantities (cf. e.g. the mean field pressure at vanishing chemical potential, Section 4.6).

4.5 Expansion for small chemical potentials

Straightforward Monte-Carlo sampling in lattice calculations is limited to zero chemical potential due to the sign problem. One way to still obtain information about the pressure at nonzero chemical potential μ is to investigate the coefficients of the Taylor series with respect to μ [All02]

$$\frac{p(T,\mu)}{T^4} = \sum_{n=0}^{\infty} c_n(T) \left(\frac{\mu}{T}\right)^n$$

$$\text{with} \quad c_n(T) = \frac{T^{n-4}}{n!} \left(\frac{\partial^n p}{\partial \mu^n}\right)\bigg|_{\mu=0} \quad (4.2)$$

also dubbed susceptibilities, where c_0 is the pressure at $\mu = 0$ and c_2 and c_4 are the next nonzero coefficients[4].

For the eQPM, the coefficients are

---

[4]Within the quasiparticle model all odd-numbered coefficients vanish in the limit $\mu \to 0$ due to sign changes of the respective derivatives of statistical distribution functions. This can be shown be a general feature due to the symmetry of the considered observables with respect to changing μ to $-\mu$ [All02].

4.5 Expansion for small chemical potentials

$$c_2^{eQP} = \frac{d_q}{2\pi^2 T^3} \int_0^\infty dk\, k^2\, ef^2,$$

$$c_4^{eQP} = \frac{d_q}{24\pi^2 T^3} \int_0^\infty dk\, k^2\, ef^4 \left[3T(1-e^2)\left.\frac{\partial^2 \omega_{\text{TL}}}{\partial \mu^2}\right|_{\mu=0} + e^2 - 4e + 1 \right] \quad (4.3)$$

with $e = e^\pm|_{\mu=0}$ and $f = f_\pm|_{\mu=0}$ and

$$\left.\frac{\partial^2 \omega_{\text{TL}}}{\partial \mu^2}\right|_{\mu=0} = \frac{1}{3\omega_{\text{TL}}} \left[\frac{G^2}{\pi^2} + \frac{3\tilde{m}_{q,\infty}}{\pi^2 T}\sqrt{\frac{G^2}{6}} + \left(\frac{3\tilde{m}_{q,\infty} T}{2\sqrt{6G^2}} + \frac{T^2}{2} \right) \left.\frac{\partial^2 G^2}{\partial \mu^2}\right|_{\mu=0} \right]. \quad (4.4)$$

The second derivative of the effective coupling with respect to the chemical potential $\partial^2 G^2/\partial \mu^2$ can be found by differentiating the flow equation with respect to μ. In the limit of vanishing chemical potential this leads to

$$\left.\frac{\partial^2 G^2}{\partial \mu^2}\right|_{\mu=0} = \left. \left(\frac{1}{a_\mu} \frac{\partial b}{\partial \mu} - \frac{1}{a_\mu} \frac{\partial a_T}{\partial \mu} \frac{\partial G^2}{\partial T} \right) \right|_{\mu=0} \quad (4.5)$$

with the derivatives of the flow equation coefficients given in Appendix D. The first term within c_4^{eQP} featuring the second derivative of the dispersion relation with respect to the chemical potential is responsible for a peak structure of this coefficient.

Only few lattice calculations of the c_i are available. Previous comparisons [Blu05a] have been performed for $N_f = 2+0$ lattice results with small temporal extent ($N_\tau = 4$) and unphysical quark masses [All03] (p4 lattice action). Deviations of about 20 percent were found between the then-current lattice results for the pressure [Pei00] and the c_i. To check the flexibility of the model, the latter was readjusted to describe the pressure Taylor coefficients, yielding a good description.

More recently, lattice calculations for $N_f = 2+1$ with almost physical quark masses but still small lattice spacings $N_\tau = 4$ and 6 have been published by a different group [DeT10] (asqtad lattice action). From our past experience, lattice results for $N_\tau = 6$, especially for the asqtad action, still show large discretization errors and are therefore unsuited for a strict numerical comparison with our continuum model. One indication for the latter presents the extremely high second Taylor coefficient, which for $N_\tau = 4$ even surpasses the Stefan-Boltzmann limit, from [DeT10] in comparison to [All03, All05].

To gain some insight into the extrapolation procedure a qualitative comparison is still useful. The left panel of Fig. 4.12 shows the most recent lattice results for c_2 and c_4 from the the two groups in comparison to the pressure coefficients following from the eQPM using the parametrization for the results from [Baz09], where the critical temperature estimate agrees with [DeT10]. Indeed, the results agree qualitatively, with c_2 rising and c_4 falling monotonically

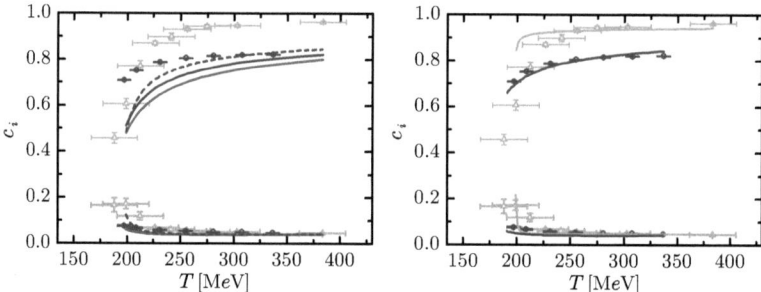

Figure 4.12: Left panel: The Taylor coefficients c_2 (upper part) and c_4 (lower part) of the pressure are shown for the eQPM adjusted to the results from [Baz09] for the p4 action with $N_\tau = 6$ (blue dashed curve) and $N_\tau = 8$ (solid blue curve) as well as the asqtad action with $N_\tau = 8$ (solid red curve) for values above $T_c = 190$ MeV. They are contrasted to the lattice results from [DeT10] for $N_f = 2+1$ using the asqtad action with $N_\tau = 6$ (yellow symbols). For reference, also the $N_f = 2+0$ lattice results using the p4 action with $N_\tau = 4$ [All05] are given (blue symbols). Right panel: Readjustment of the eQPM to the Taylor coefficients c_2 and c_4 of the lattice results.

with the temperature.

Also investigating the flexibility of the model versus the more recent lattice results yields an inferior result (cf. Fig. 4.12). This is due to the very large estimates for c_2 which approach the Stefan-Boltzmann limit very rapidly. The eQPM is unable to simultaneously describe at the same time, the values of c_2 in the smooth incline from T_c onwards and the very large asymptotic values of c_2. The fit[5] prefers the latter for their smaller error bars. However, due to the current status of c_i lattice results, it remains unclear whether the predicted large c_2 are either due to discretization effects or indeed contradict the eQPM which favors smaller c_2. At least a naive continuum extrapolation of the $N_\tau = 4$ and $N_\tau = 6$ lattice results towards the continuum limit $N_\tau = \infty$ indicates a considerably reduced c_2.

In addition to the case of one independent chemical potential $\mu = \mu_u = \mu_d$ with $\mu_s = 0$ under consideration here, situations with two and more independent chemical potentials are also investigated on the lattice. In such cases c_2 and c_4 are referred to as unmixed coefficients or flavor diagonal susceptibilities, sometimes denoted as c_{20} and c_{40}, whereas off-diagonal susceptibilities appear from the mixed derivatives with respect to the the different chemical potentials. A modified version of the eQPM has been tested against the lattice data for the off-diagonal susceptibilities from [Gav05a, Gav05b] in [Blu08b].

The check of the eQPM against lattice calculations for imaginary chemical potential [DEl04, DEl07] has been performed successfully in [Blu08a].

---

[5]The eQPM parameters obtained from the simultaneous adjustment to c_2 and c_4 with $\chi^2 = (\chi^2_{c_2} + \chi^2_{c_4})/2$ are $T_s = 147$ MeV, $\lambda = 17$ MeV using [All05] lattice results and $T_s = 198$ MeV, $\lambda = 0.07$ MeV using [DeT10] lattice results.

4.6 Results at small temperatures

The extrapolation procedure ends with the characteristic curves touching the μ-axis at $T = 0$. However, as visible from the results in Section 4.3, all thermodynamic quantities (except for the entropy density which vanishes in this limit) are almost constant as functions of T in the interval $T\epsilon[0, 100]$ MeV for fixed μ. We therefore investigate the results for $T = 25$ MeV, allowing for a easier comparison to the HTL QPM in Chapter 5.

The eQPM pressure, net particle and energy density at $T = 25$ MeV are shown – appropriately scaled – in Fig. 5.12 as functions of the chemical potential $\mu = \mu_q$ (grey curves) in comparison to the corresponding HTL quantities (green curves) for both the adjustment to [Bor10b] lattice data and lattice results from [Baz09] for the p4 action with $N_\tau = 8$. All turn out to be monotonically increasing with growing chemical potential. While the eQPM net particle and energy densities are strictly positive, the eQPM pressure turns negative at the lowest μ reached by the characteristic curves. This gives rise to a small region of negative pressure which in Fig. 4.4 is covered by the pseudocritical characteristic ending at a larger chemical potential. We will address this region thoroughly in Section 5.4.

Since the statistical distribution function of bosons $n_\text{B} = [f_+ + f_-]_g/2$ vanishes in the limit $T \to 0$ one might expect, from Eqs. (4.1), a constant gluon mean field pressure B_g at vanishing temperature. However, it turns out to be a non-constant function of the chemical potential (not shown). This is due to to the fact that thermodynamic self-consistency is ensured for the overall mean field pressure B only, yet not for the particular contributions B_i (cf. Sections 2.9 and 4.4).

The integration of the individual mean field pressure contributions via the characteristics amount to a specific choice for the integration paths which leads to a specific distribution of contributions B_i to the overall mean field pressure B (in particular with $B_g \neq$ const). Choosing a different integration path, e.g. along the μ-axis, leads to an entirely different distribution, e.g. with $B_g =$ const. This does not impair the overall B in any way but instead, in showing the same result for completely different integrations, underlines the overall consistency of the model.

4.7 The explicit dependence of the asymptotic masses on the chemical potential

It is instructive to consider the case of modified thermal masses in the asymptotic dispersion relation, where

$$m_{q,\infty}^2 \sim \left(T^2 + \frac{\mu^2}{\chi^2}\right) G^2 \quad \text{and} \quad m_{g,\infty}^2 \sim \left(\frac{C_b}{6}T^2 + \frac{\mu^2}{4\chi^2}\right) G^2 \qquad (4.6)$$

so that for $\chi \to \infty$ the dependence of the thermal masses on the chemical potential μ is eliminated and $\partial m_{i,\infty}^2/\partial \mu|_{G^2} = 0$.

Dropping also the restmasses[6] $m_{i,0}$ and thus reverting to simple asymptotic dispersion relations and masses, all of the latter vanish in the limit $T \to 0$. Furthermore, the derivative of the asymptotic masses $m_{i,\infty}^2$ with respect to the chemical potential at constant G^2 vanishes so that at $T = 0$

$$\frac{\partial m_{i,\infty}^2}{\partial \mu} = \frac{\partial m_{i,\infty}^2}{\partial \mu}\bigg|_{T,G^2} + \frac{\partial m_{i,\infty}^2}{\partial G^2}\frac{\partial G^2}{\partial \mu} \quad (4.7)$$

is expected to vanish too.

If, indeed, the mass derivatives vanish, all mean field pressure contributions B_i^{eQP} should, according to Eq. (4.1), turn constant at $T = 0$, leading to a constant overall mean field pressure $B^{eQP} = B_{\text{const}}$. For the gluons this is already required by properties of the Bose distribution, cf. Section 4.6, which also leads to $p_g^{eQP}(T = 0) = 0$. From the expansion of the quasiparticle partial pressure (3.18) and neglecting heavy quark flavors one then has for the case $\chi \to \infty$ and $T \to 0$

$$p^{eQP} = \frac{d_q}{24\pi^2}\mu^4 - B_{\text{const}} \quad (4.8)$$

which corresponds to a simple bag model pressure.

If, on the other hand, one integrates the mean field pressure along a characteristic curve, one finds from (2.55) with $\Pi_i \to m_{i,\infty}^2$ that, due to $\partial m_{i,\infty}^2/\partial \mu|_{G^2} = 0$

$$a_T \frac{\partial m_{i,\infty}^2}{\partial T}\bigg|_{G^2} = -b\frac{\partial m_{i,\infty}^2}{\partial G^2} \quad (4.9)$$

so that Eq. (2.62)

$$B^{eQP} = B_0^{eQP} - \int \mathrm{d}x \sum_i \left(\underbrace{a_T \frac{\partial m_{i,\infty}^2}{\partial T}\bigg|_{G^2} + b\frac{\partial m_{i,\infty}^2}{\partial G^2}}_{0} + \underbrace{a_\mu \frac{\partial m_{i,\infty}^2}{\partial \mu}\bigg|_{G^2}}_{0} \right) \frac{\partial p_i^{eQP}}{\partial m_{i,\infty}^2} = B_0^{eQP}, \quad (4.10)$$

i.e. the mean field pressure $B_0^{eQP} = B^{eQP}(T_0)$ at the emergence temperature of the characteristic curve is transported along the latter to the μ-axis, so that $B^{eQP}(T = 0, \mu = \mu_f) = B^{eQP}(T = T_0, \mu = 0)$. This is, however, not a constant.

The discrepancy is easily solved, as the assumption of vanishing derivatives of the asymptotic masses with respect to the chemical potential is not realized. The solution of the flow equation of this modified eQPM yields a diverging effective coupling G^2 in the limit $T \to 0$. Therefore, the derivative of the coupling with respect to the chemical potential also diverges. As a consequence, the product of diverging $\partial G^2/\partial \mu$ and converging $\partial m_{i,\infty}^2/\partial G^2$ (which is not a function of the

---

[6]The considerations may as well be performed for finite restmasses, yielding no change for the mean field pressure, however introducing additional terms to the pressure expansion. For brevity, the simpler case is outlined.

diverging G^2) yields a finite result along any characteristic in the limit $T \to 0$. Accordingly, there are finite contributions to the μ-integral in the partial mean field pressures (cf. Eq. (4.1)) resulting in a non-constant overall mean field pressure equal to the one obtained by integration along the characteristic curves.

It is very intriguing that the thermodynamic quantities of the eQPM with and without the modification turn out quite comparable. Depending on the quantity, differences of about 10 to 20 percent were found. Even the paths of the characteristic curves turn out to be very similar. This supports the work done by M. Bluhm on multiple (small) independent chemical potentials, where the dependence of the thermal masses on the chemical potential had to be dropped in order to preserve self-consistency [Blu08b].

4.8 Wrap-up

Concluding the chapter, we may summarize that the established eQPM is able to describe the most recent lattice results. As for past adjustments, the extrapolation of state quantities is free of irregularities; crossing characteristics can be dealt with in the usual fashion. The expected semi-quantitative agreement of the eQPM extrapolation procedure with the lattice predictions for the pressure susceptibilities was confirmed and self-consistency of the eQPM verified.

Finally, it was shown that the approach to neglect the explicit dependence of the asymptotic masses on the chemical potential in order to preserve thermodynamic self-consistency for the case of multiple independent chemical potentials, as adopted in [Blu08b], introduces only small deviations.

5 Equation of state for heavy-ion collider experiments

We now return to the HTL QPM in order to compare it to the established eQPM and assess its improvements. In doing so, the state quantities of the two models at zero and nonzero chemical potential as well as the pressure Taylor coefficients are contrasted. Particular attention is paid to the role of the collective modes. We then investigate to possibility to connect the HTL QPM to the hadron resonance gas enabling us to construct an EOS for heavy-ion collider experiments at large net baryon densities. Finally, the application of the HTL quasiparticle model EOS in the calculation of gluon condensates as input for sum rule studies is presented.

5.1 Comparison with lattice results

As for the eQPM (cf. Section 4.2), the HTL QPM parameters have, in a first step, to be adjusted to lattice results. This is carried out as for the former with resulting parameters shown in Tab. 5.1. The underlying lattice data for the scaled interaction measure $(e - 3p)/T^4$ as well as the results for the HTL QPM using the adjusted parameters are shown in Fig. 5.1. It is quite remarkable that, apart from the result for the older lattice data with $N_\tau = 6$, the HTL interaction measure resembles the eQPM trace anomaly despite the fact that the former does not account for quark restmasses. This may be viewed as a (first) sign of consistency between the two quasiparticle models.

Even from the change in χ^2/dof no determination can be made whether the description of the one model is better than the other, as it increases for the fit to [Bor10b] lattice results and marginally decreases for the adjustment to [Baz09] lattice data. It is noteworthy that the parameters λ and T_s seem to be almost fixed for each lattice result as both keep roughly the same values when switching from eQPM to the HTL QPM (cf. Tab. 4.1) – the exception being T_s in the stout case. In contrast, the constant B_c decreases noticeably for all four cases, signifying that the additional modes of the HTL QPM provide an overall negative pressure contribution as B_c is the integration constant of the mean field pressure being subtracted from p (cf. Eqs. (2.44) and (2.61)).

From the almost equal interaction measure it is no surprise that all other state quantities from the lattice are matched equally well as for the eQPM model (cf. right panel of Fig. 5.1). Both

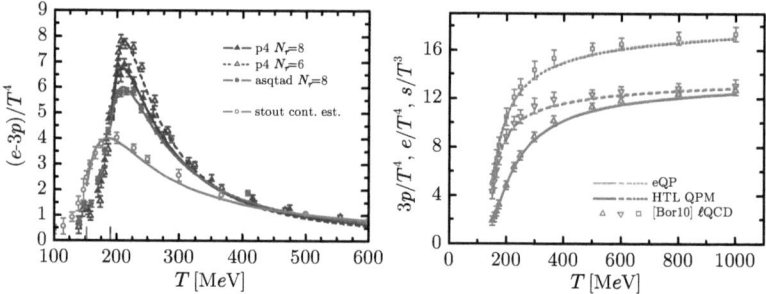

Figure 5.1: Left panel: Comparison of the interaction measure of the HTL QPM to the lattice QCD data (symbols) for lattice actions p4 (blue), asqtad (red) and stout (green) and lattice spacings N_τ from [Che07, Baz09, Bor10b]. Solid (dashed) curves are for $N_\tau = 8$ (6). The statistical errors from the lattice are indicated. Right panel: Comparison of scaled pressure $3p/T^4$ (upright triangles, solid lines), scaled energy density e/T^4 (reversed triangles, dashed lines) and scaled entropy density s/T^3 (circles, dotted lines) derived from the interaction measure on the lattice (symbols) and via the HTL QPM (curves) adjusted to lattice results from [Bor10b]. Statistical errors from the lattice are indicated where available. Both panels: For reference, the eQPM results (cf. Figs. 4.1 and 4.2) are shown as grey curves which, however, are barely visible as they are beneath the HTL QPM curves.

action	N_τ	$p(T_c)/T_c^4$	T_s [MeV]	λ [MeV]	B_c	χ^2/dof
p4	6	0.58	170	16	$-(218 \text{ MeV})^4$	2.50
p4	8	0.70	149	26	$-(101 \text{ MeV})^4$	1.18
asqtad	8	0.76	131	36	$-(112 \text{ MeV})^4$	3.23
stout	∞ (est.)	0.63	31	85	$-(144 \text{ MeV})^4$	1.15

Table 5.1: Parameters of the HTL QPM as results of the adjustment of the interaction measure to the lattice QCD results [Che07, Baz09, Bor10b] with the pressure $p(T_c)$ being fixed to the lattice pressure p_{lattice} at T_c (cf. Section 2.12).

models are therefore equivalent at vanishing chemical potential and one should, in calculations restricted to $\mu = 0$, rather use the simple eQPM than the more sophisticated HTL QPM which requires integrations in energy and momentum space.

5.2 Contributions of collective modes

As argued in Chapter 3, the contributions of the collective excitations to the net quark density as well as the (partial) pressure and entropy density are negative. Here, these arguments are quantified for the adjustment of the HTL QPM to lattice results from [Bor10b].

Fig. 5.2 shows the contributions of plasmons, plasminos and non-collective modes to the pressure, entropy and energy density as well as the interaction measure as functions of the temperature at vanishing chemical potential. The largest contributions from plasmons and plasminos to the entropy density and pressure[1] are found close to the phase transition. While the

---
[1]Here, we consider the pressure contributions $p_{(i)} = p_i - B_i$ in order to exclude effects from the

5.2 Contributions of collective modes

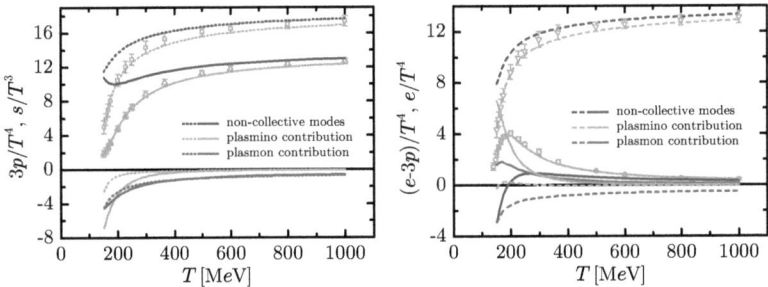

Figure 5.2: The contributions of plasmons (light blue curves), plasminos (yellow curves) and the non-collective modes (dark green curves) to the scaled pressure $3p/T^4$ (left panel, solid curves), the entropy density (left panel, dotted curves), the energy density (right panel, dashed curves) and the interaction measure (right panel, solid curves) of the HTL QPM adjusted to [Bor10b] lattice results at $\mu = 0$ are shown. For reference, the lattice results and the overall quantities are shown (light grey symbols and curve, respectively; cf. Fig. 5.1)

plasmino contribution already vanishes at rather low temperatures, the plasmon contribution gives noticeable contributions up to very high temperatures. While the absolute value of the plasmon contribution is larger than the absolute value of the plasmino contribution for the entropy at all temperatures, it is smaller than the latter for the pressure at small temperatures close to T_c and larger for $T > 195$ MeV only.

The result for the energy density follows directly from the Gibbs relation with $n_q(\mu = 0) \equiv 0$. Interestingly, the plasmino contributions from the pressure and the entropy density almost cancel each other, leading to a very small contribution to the energy density, even alternating in sign. As the overall size of the entropy density plasmon contribution is larger than the plasmon contribution to the pressure, the energy density plasmon contribution remains negative and also quite sizeable up to large values of T.

The results for the interaction measure are determined via $(e - 3p)/T^4 = s/T^3 - 4p/T^4$ (cf. Eqs. (2.67)). Here, the plasmon contribution from the pressure is enhanced by a factor of four in comparison to the energy density leading to a positive contribution of the longitudinal gluons to the trace anomaly. The same is true for the plasminos, where the contribution turns out quite large, especially close to T_c. The contributions from both collective modes vanish more rapidly for the interaction measure than for the other state variables with the plasmino contribution still going to zero faster than the plasmon contribution. It is due to the contributions of the collective modes to $e - 3p$ turning out positive, that the contribution from the non-collective modes is determined to be partially negative close to T_c. The sign change occurs at about the temperature where the contributions from plasminos and plasmons are equal.

---

overall pressure integration constant B_c (cf. Eqs. (2.51) and (2.52)). Consequently, the sum of all contributions does not add up to p but $p + B_c$.

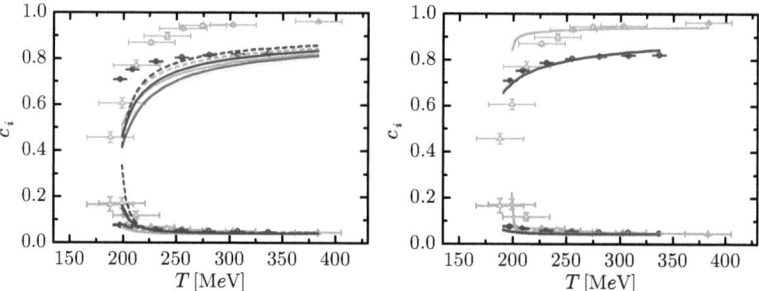

Figure 5.3: Left panel: The Taylor coefficients c_2 (upper part) and c_4 (lower part) of the pressure are shown for the HTL QPM (colored curves) and the eQPM (greyscale curves) adjusted to the results from [Baz09] for the p4 action with $N_\tau = 6$ (blue dashed/grey curves) and $N_\tau = 8$ (solid blue/grey curves) as well as the asqtad action with $N_\tau = 8$ (solid red/light grey curves) for values above $T_c = 190$ MeV. They are contrasted to the lattice results from [DeT10] for $N_f = 2 + 1$ using the asqtad action with $N_\tau = 6$ (yellow symbols). For reference, also the $N_f = 2 + 0$ lattice results using the p4 action with $N_\tau = 4$ [All05] are given (blue symbols). Right panel: Readjustment of the HTL QPM to the Taylor coefficients c_2 and c_4 of the lattice results.

While, per se, a negative overall particle density or pressure is unphysical, the negative contributions rather describe the effect of the collective excitations in the QGP. For instance for the entropy density: as medium effects indicate correlations between the gas-like constituents of the eQPM plasma, taking them into account causes a decrease in overall entropy density. In the same way, also the pressure and the net particle density are decreased by the introduction of the collective modes.

5.3 Pressure susceptibilities

The Taylor coefficients c_i of the pressure (4.2) may be used to compare the extrapolation procedure of the HTL QPM with the extrapolation procedure of the eQPM and – to some extent (cf. Section 4.5) – the prediction by lattice calculations. The expressions for the susceptibilities of the HTL QPM are given in Appendix D.

As visible from Figs. 5.1 (the pressure at $\mu = 0$ corresponds to c_0) and 5.3 both predictions for the pressure at $\mu \gtrsim 0$ are quite similar. The pressure at vanishing chemical potential c_0 is virtually identical and the c_2 of the HTL QPM are only slightly larger than their eQPM equivalents. A difference is, however, perceptible in c_4, where a more pronounced increase for the temperature decreasing towards T_c is seen. This corresponds to the qualitative prediction of the lattice results which show a peak of c_4 at T_c. In this way one may conclude that the HTL QPM Taylor coefficients give a better (qualitative) description of lattice susceptibilities than the c_i found for the eQPM.

5.4 Extrapolation to nonzero chemical potential

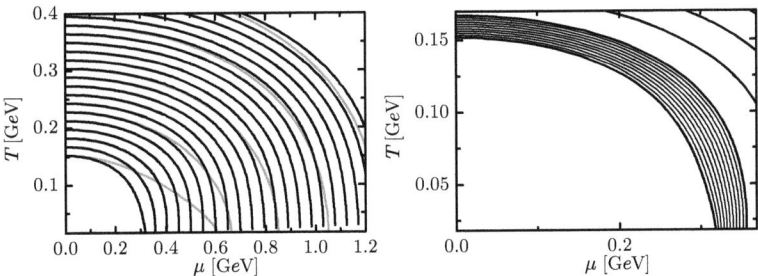

Figure 5.4: Left panel: Some characteristic curves $(T(x), \mu(x))$ of the flow equation (2.56), emerging from the temperature interval $T_0\epsilon[152, 450]$ MeV, are shown for the HTL QPM adjusted [Bor10b] lattice results (black curves). For reference the grey curves indicate characteristics of the eQPM (cf. Fig. 4.3). Right panel: Zoom into the left panel with additional characteristics in the interval $T_0\epsilon[152, 167]$ MeV close to the transition which is prone to crossing characteristics in the eQPM. In the HTL QPM no crossings of characteristic curves appear.

A readjustment of the model parameters as performed for the eQPM is also possible[2], leading to remarkably similar results (cf. right panel of Fig. 5.3). Taking into account that the fit of the HTL and the eQPM interaction measure to the lattice results were notably similar, too (Fig. 5.1), one may conclude that – at least at $\mu \approx 0$ – the additional collective modes, Landau damping and the use of the full dispersion relation do not affect the flexibility of the model. That is, neither in a positive nor negative way.

On the one hand, this represents a second hint, that indeed the approximations put forward to derive the effective from the HTL QPM are good approximations. On the other hand, this also means that the positive results found for the eQPM in numerous publications can, for $\mu \approx 0$, be assumed to be valid for the HTL QPM as well. This includes comparisons with actual experimental results via hydrodynamic calculations [BKS07b] as well as lattice results for imaginary chemical potential [Blu08a] and off-diagonal susceptibilities [Blu08b].

5.4 Extrapolation to nonzero chemical potential

As claimed [Rom04] and confirmed [Sch07, Sch08a] for lattice results from [Kar07, Che07], the characteristic curves of the HTL quasiparticle model show no crossings as encountered for the eQPM. This also holds for the adjustment to the lattice results from [Bor10b] (cf. Fig. 5.4) and [Baz09] (not shown) and seems to be a general feature.

Using the hint from [Blu04] that the crossing of characteristics in the eQPM is due to the effective coupling G^2 being too large near the pseudocritical temperature, this can be explained. Since the eQPM or quasiparticle contribution to the HTL QPM entropy density increases

---

[2]The model parameters obtained from the simultaneous adjustment to c_2 and c_4 with $\chi^2 = (\chi^2_{c_2} + \chi^2_{c_4})/2$ are $T_s = 135$ MeV, $\lambda = 22$ MeV using [All05] lattice results and $T_s = 197$ MeV, $\lambda = 0.19$ MeV using [DeT10] lattice results.

with decreasing mass parameters m_D^2 and $\hat{M}^2$ which are proportional to G^2T^2 at $\mu = 0$, the crossings would therefore disappear for a larger quasiparticle contribution. One way to allow for a larger quasiparticle contribution to the entropy density is to take into account collective modes. Due to their negative contributions, the model parameters have to change, increasing the quasiparticle contribution (as visible Fig. 5.2), in order to still describe the lattice data. With the resulting decrease of the effective coupling G^2 the crossings then disappear.

Another degree of freedom suitable to decrease overall entropy density is a finite quasiparticle width [Pes04, Pes05]. Depending on the size of and the way the widths are introduced into the QPM, the crossings can thus be removed [Sch07].

Figures 5.5 through 5.11 show the thermodynamic quantities as results of the extrapolation procedure using the lattice results from [Bor10b] as boundary values within the T-μ-plane as functions of the temperature at several values of the chemical potential and additionally as contour plots. As for the eQPM, all considered state variables of the HTL QPM, pressure, interaction measure, entropy density, net quark density and energy density, increase with rising chemical potential at constant temperature. The mean field pressure B increases with μ at smaller temperatures while it starts to decrease if considering larger temperatures. The effective coupling G^2 decreases with increasing chemical potential at constant temperature.

In the figures, the scaling with powers of T, which is customary at $\mu = 0$, is kept for the cuts along constant $\mu = 0$, 0.2 GeV and 0.3 GeV. As visible from the contour plots, the resulting divergences close to $T = 0$ are a solely due to this choice of the scaling and can be safely ignored. As for the eQPM, no irregularities arise from the extrapolation and the thermodynamic quantities at $T = 0$ as functions of μ resemble the ones as functions of T at $\mu = 0$ with specific scalings and shifts.

Focusing on the contour plots, the most noteable fact, however, is that the HTL QPM results for the pressure, the entropy density, the net quark density and the energy density all are very close to the eQPM results. This is especially striking for the former two, where almost no differences are noticeable, while some small differences are visible for the latter two.

On the other hand, this is not the case for the interaction measure $e - 3p$ (Fig. 5.8) and the pure mean field pressure $B - B_0$ (Fig. 5.10) which differ visibly from the eQPM results and instead, as for the eQPM, rather stay (almost) constant along the path of the new characteristic curves. This illuminates the role of the characteristics within the QPM and connects the two quantities in an intricate way. The fact, that the small changes in the other state quantities cause a rather large shift in the interaction measure indicates the sensible nature of this quantity.

Due to the only small changes in the other thermodynamic quantities most properties from the eQPM are preserved. The lines of constant pressure and the lines of constant energy density still resemble each other (cf. Fig. 5.5 and 5.7), however, not as closely as for the eQPM. The limits $n_q(\mu \to 0) = 0$ (Fig. 5.9) and $s(T \to 0) = 0$ (Fig. 5.6) still hold, as required from general thermodynamics.

In Section 4.3 we found that, for the eQPM at the lowest chemical potentials reached

5.4 Extrapolation to nonzero chemical potential

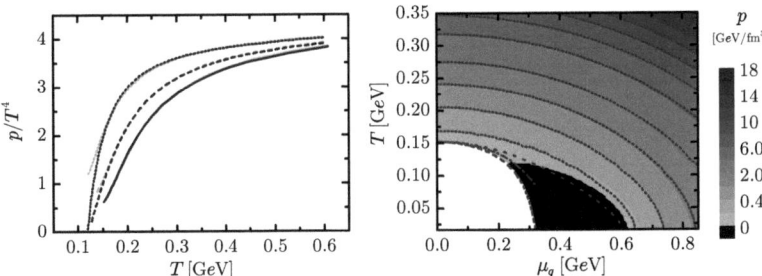

Figure 5.5: Left panel: The scaled overall pressure p/T^4 of the HTL QPM adjusted to [Bor10b] lattice results (dark red curves) is shown as a function of the temperature T for chemical potentials $\mu = 0$ (solid curve), 0.2 GeV (dashed curve) and 0.3 GeV (dotted curve). For reference, the eQPM results, matched to the same lattice results at $\mu = 0$, are shown as well (grey curves, cf. Fig. 4.4). Right panel: Contour plot of the pressure p of the HTL QPM contrasted to the lines of constant pressure of the eQPM (dotted curves). The black area indicates the region of negative overall pressure. For reference, the pseudocritical characteristics of the HTL QPM (dashed curve) and the eQPM (wide dotted curve) emerging from $T_0 = T_c$ are shown as well as the lattice prediction for the chiral phase transition line [Kac10].

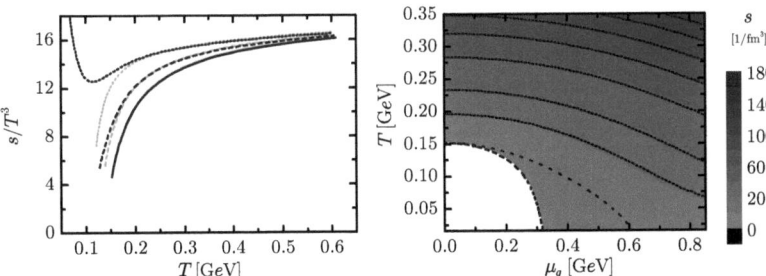

Figure 5.6: As Fig. 5.5 but for the scaled overall entropy density s/T^3. For the eQPM results (left panel: grey curves, right panel: dotted curves) cf. Fig. 4.6.

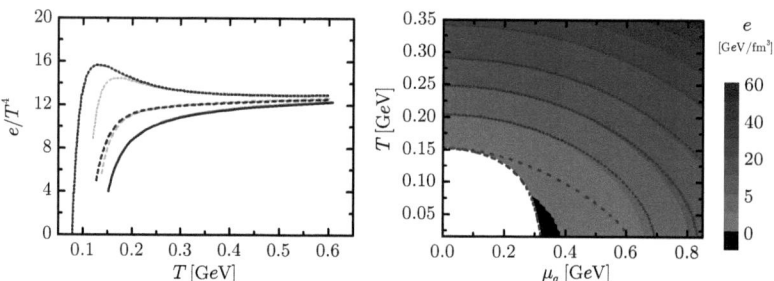

Figure 5.7: As Fig. 5.5 but for the scaled overall energy density e/T^4. For the eQPM results (left panel: grey curves, right panel: dotted curves) cf. Fig. 4.8.

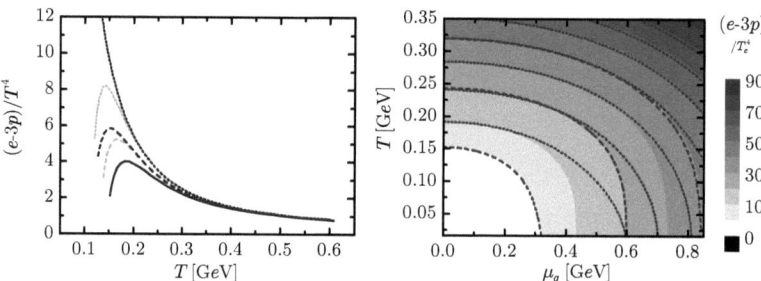

Figure 5.8: As Fig. 5.5 but for the scaled overall interaction measure $(e-3p)/T^4$. For the eQPM results (left panel: grey curves, right panel: dotted curves) cf. Fig. 4.5. The characteristic curves of the HTL QPM emerging from $T_0/T_c = (1.0, 1.6, 2.1)$ are shown as dashed curves in the right panel.

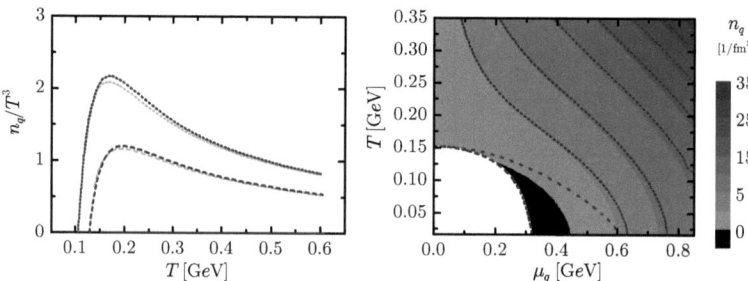

Figure 5.9: As Fig. 5.5 but for the scaled overall net quark density n/T^3. For the eQPM results (left panel: grey curves, right panel: dotted curves) cf. Fig. 4.7.

5.4 Extrapolation to nonzero chemical potential

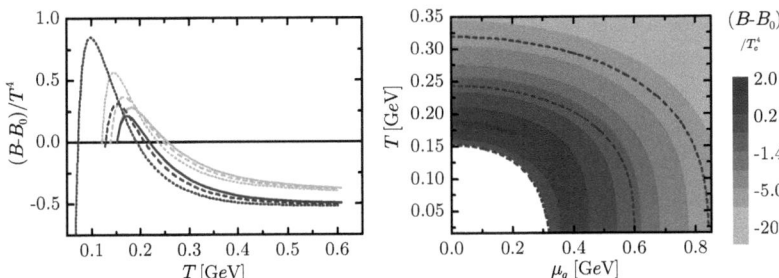

Figure 5.10: As Fig. 5.5 but for the scaled mean field pressure B/T^4. For the eQPM results (grey curves) cf. Fig. 4.9. The characteristic curves emerging from $T_0/T_c = (1.0, 1.6, 2.1)$ are shown as dashed curves in the right panel.

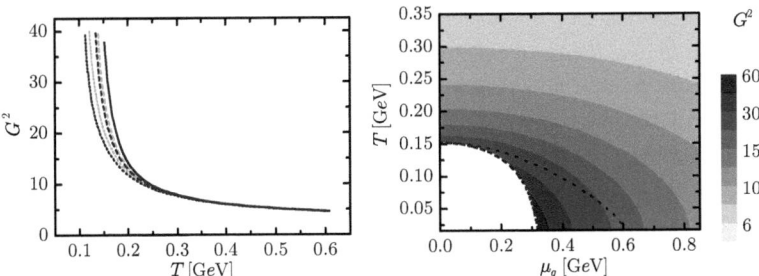

Figure 5.11: As Fig. 5.5 but for the effective coupling G^2. For the eQPM results (grey curves) cf. Fig. 4.10.

by the characteristic curves, the pressure turned negative. As visible from the Fig. 5.5, the characteristics of the HTL QPM cover an enlarged area reaching down to small chemical potentials $\mu \gtrsim 0.3$ GeV where the pressure is found to be negative. This is only partly accounted for by the similarity of both models with the HTL QPM extending the eQPM results to lower chemical potentials. As argued in Section 3.3 and shown explicitly for $\mu = 0$ in Fig. 5.2 it is due to the negative plasmino and plasmon contributions that the region of negative pressure extends even somewhat above the pseudocritical curve of the eQPM.

The large differences in the eQPM and HTL QPM pseudocritical characteristics show that the concept of using the characteristic emerging from $T_0 = T_c$ as some indication for the transition line (cf. Section 2.12) is somewhat imprecise for large μ, and it is advisable to consider other concepts to better approximate it (cf. Section 5.5). Comparing the pseudocritical characteristics with predictions for the chiral phase transition line at low chemical potential from the Bielefeld group [Kac10] we find that they are at least in the right ballpark.

In any case, the region of negative pressure, which contains the smaller regions of negative net quark and energy density, indicates that the QGP with properties predicted by the HTL QPM is not a stable state of matter in this region. As any other state of matter with non-negative pressure is preferred the line of vanishing pressure represents the maximum extent of the QGP, however it is most likely, that the transition occurs at pressures larger than zero. This transition from the QGP to the other state of matter is discussed in Section 5.5.

Fig. 5.12 shows the pressure as well as the net particle and energy density at almost vanishing temperature $T = 25$ MeV for adjustments to [Bor10b] lattice data and lattice results for the p4 action with $N_\tau = 8$ [Baz09]. The difference to the results at $T = 0$ is about 2 to 5 percent depending on the quantity. Comparing the eQPM and the HTL QPM, only small differences in the pressure are visible while the variations in the two derived quantities (cf. Eqs. (2.47) and (6.18)) are slightly larger. Considering the large modifications from the HTL QPM to the effective QPM (cf. Section 4.1), the small extent of the changes is quite remarkable.

In view of the similar results of both models (recall also the equivalent flexibilities in the description of lattice results, cf. Sections 5.1 and 5.3) and the much more accessible analytic structure of the eQPM, the latter is a much more suitable choice for explorative studies without requirement of absolute numerical accuracy (as e.g. performed in Chapter 6). The crossing of characteristics for the eQPM can be put up with as long as it is dealt with in a proper manner (cf. Section 2.10).

5.5 Equation of state for SPS

We now focus on the EOS needed for the hydrodynamic description of the fireball phase of a relativistic heavy-ion collision. Assuming adiabatic cooling the EOS is needed along curves of constant entropy per particle s/n, dubbed adiabates or isentropes.

The left panel of Fig. 5.13 shows the lines of constant entropy per particle of the HTL QPM

5.5 Equation of state for SPS

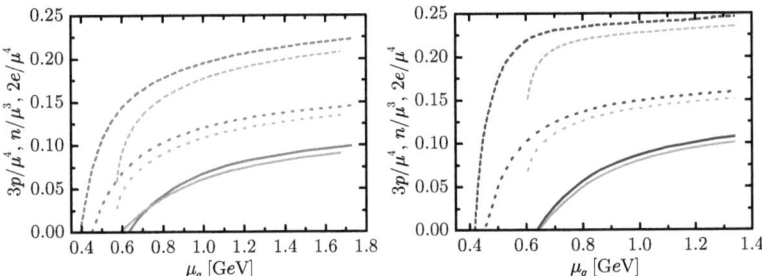

Figure 5.12: The scaled thermodynamic state variables p/μ^4 (solid curve, scaled by factor 3), n/μ^3 (dotted curve) and e/μ^4 (dashed curve, scaled by factor 2) are shown as functions of the chemical potential $\mu = \mu_q$ at constant temperature $T = 25$ MeV. The difference to the results at $T = 0$ is about 2 to 5 percent depending on the quantity. The colored lines correspond to results for the HTL QPM adjusted to [Bor10b] lattice data (left panel, green curves) and results from [Baz09] for the p4 action with $N_\tau = 8$ (right panel, blue curves).

on top of the pressure contour plot. For small values of $s/n_q \lesssim 3.5$ the trajectories $(T(p), \mu(p))$ along the adiabates move from regions of high temperature and chemical potential to regions of smaller T and μ with decreasing pressure. For higher entropies per particle the trajectories exhibit a bending, causing the chemical potential to increase with decreasing pressure while the temperature continues to drop[3]. Along the adiabates both net particle and entropy density decrease. Trajectories with $s/n_q \lesssim 27$ enter the region of negative pressure, where any other state of matter with non-negative pressure is preferred to the QGP described by the QPM. The transition from the quark and gluon degrees of freedom to this other kind of matter is to occur outside of this region.

In order to provide an EOS for hydrodynamic calculations at e.g. SPS and FAIR the EOS of the HTL QPM which is valid for the deconfined phase has to be connected to an EOS for the confined phase. The hadron resonance gas (HRG) is a rather simple yet quite successful model for the confined phase of the QCD phase diagram[4]. It is, therefore, a logical first choice as hadronic counterpart to our EOS.

As visible from the right panel of Fig. 5.13, the pressures of both phases intersect at $\mu = 0$ making a construction between the two unproblematic. As a consequence, the lattice groups use the HRG as a benchmark for their results at vanishing chemical potential [Bor10a, Bor10b,

---

[3]While this feature is also visible for the eQP isentropes, it is not as pronounced (cf. Fig. 5.14). This is partially due to the enlarged region of negative net particle density of the HTL QPM. At the curve $n_q = 0$ the entropy per particle diverges. Consequently, it marks a barrier the isentropes circumvent by bending back more strongly.

[4]The HRG goes back to a theorem [DMB69] allowing the calculation of state variables of an interacting system via separation of non-interacting partition function and a piece containing all the interactions. For the latter, it can be shown that, in the thermodynamic limit, attractive and repulsive forces almost exactly cancel [Ven92, Bor10b]. It is therefore a good approximation to consider hadronic matter as as gas of non-interacting free hadrons and resonances.

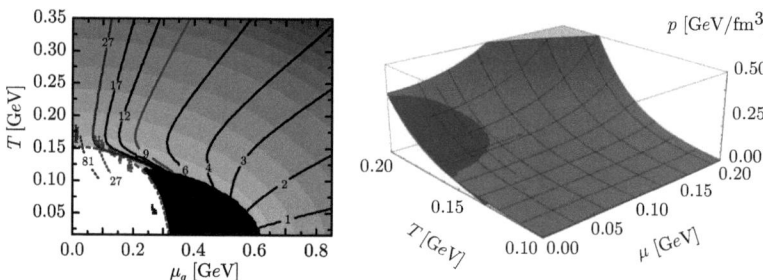

Figure 5.13: Left panel: Isentropes of the HTL QPM for several values of s/n_q (solid curves) and the hadron gas model for two corresponding values of s/n_{Ba} (dotted curves) are shown on top of the pressure contour plot (Fig. 5.5). The data points (symbols) indicate the chemical freeze-out points of RHIC (dark red), SPS (wine red) and SIS (black) from [Cle06]. Right panel: The pressure of the HTL QPM (blue plane) is contrasted to the pressure of the hadron resonance gas (red plane) as functions of temperature T and quark chemical potential $\mu = \mu_q$. The intersection line of the two planes indicates the line of equal pressure of both phases.

Huo10]. However, already at small chemical potential difficulties are met, as the extrapolation of the pressure from the lattice results [Baz09, Bor10b] as well as any QPM adjusted these begin to differ from the HRG pressure increasingly with growing chemical potential[5] (cf. right panel of Fig. 5.13).

This can be seen analytically by rewriting the expressions for the thermodynamic state variables and absorbing the explicit dependence on the chemical potential into terms $\sim \sinh(\mu/T)$ or $\sim \cosh(\mu/T)$. In the comparison of the coefficients with the HRG expressions [Kar03] the latter turn out to be larger causing the faster increase of the HRG state variables. Alternatively, the pressure susceptibilities c_i can be compared yielding, as expected, that $c_2^{HRG} > c_2^{HTL}$. At $\mu \approx 66$ MeV the line of equal pressure between the HRG and the HTL QPM ends, leaving no phase boundary for the straightforward construction of a phase transition.

On the other hand, the assumption of adiabaticity requires equal entropy per baryon in both phases. It is common practice [LR03] to set $n_q = 3n_{Ba}$ and thus $\mu = \mu_u = \mu_{Ba}/3$ to connect the quantities of hadronic and quark EOS. Assuming no change in the overall entropy we thus have $s/n_{Ba} = 3s/n_q$. From the right panel of Fig. 5.13 it is visible that, for increasing chemical potential, the isentropes of HRG and HTL QPM move further and further away from each other. In the case $s/n_{Ba} = 27$ relevant for the SPS experiment [Cle06, vHe08] the adiabates are about 100 MeV in chemical potential apart.

These issues are not unique for the case of an extrapolation of lattice results (be it using the c_i or the QPM) but have been met by other phenomenological models (e.g. in the work on

---

[5]In the past [BKS06, BKS07b], where EOS at low net baryon densities for RHIC and LHC were given based on the eQPM, we relied, therefore, on a phenomenological extension of the eQPM below T_c (cf. [Blu04a, Sch07]) rather than the HRG to describe the confined phase for small μ.

5.5 Equation of state for SPS

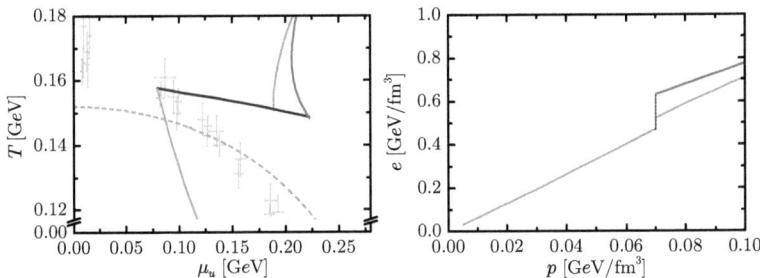

Figure 5.14: Left panel: Isentropic trajectories for $s/n_{\mathrm{Ba}} = 27$ as a result of the mixed-phase transitions (dark curves) from the HTL QPM (green curve) as well as the eQPM (grey curve) to the hadron resonance gas (yellow curve). As a consequence of the very similar pressure of the two QPM, the mixed-phase curves are virtually on top of each other. The pseudocritical characteristic of the HTL QPM is shown as grey dashed curve. The freeze-out data (grey symbols) are identical to Fig. 5.13. Right panel: The corresponding equations of state $e(p)$.

the second part of [Ste09]; private communication) as well as simple and improved bag model approaches [Bar89, Ton03, Non05]. Various modifications have been proposed, e.g. the excluded volume approach [Hag80, Ris91] or off-equilibrium descriptions of the hadronic phase [Rap01]. However, in our case, neither of those introduces a significant change of the phase boundary which would allow for a rigorous construction of a phase transition between the QPM and the confined phase along the latter.[6]

We therefore follow a phenomenological approach [Non05, vHe08] to perform a standard mixed-phase construction via a convex combination [Wal02] of thermodynamic quantities. Let the volume fraction of the matter in the hadronic phase be $x \epsilon [0, 1]$, then

$$p_M = p_H x + p_Q (1 - x) \qquad (5.1)$$

and all other quantities follow accordingly.

Since the temperature $T(x)$ and chemical potential $\mu(x)$ as functions of the volume fraction x are not fixed, but adiabaticity should hold also in the mixed phase, i.e. s_M/n_M should assume the same value as s_H/n_H in the hadronic and s_Q/n_Q in the quark-gluon phase (note that n refers to the net quark density $n_q = 3 n_{\mathrm{Ba}}$ in all cases), one condition can be imposed on the thermodynamic state quantities.

In order to, at least, provide an estimate for the EOS as if the two adiabates were connected via a phase boundary, i.e. a line $p_Q = p_H$, we may for instance require the pressure to be constant $p_M \equiv p_H(T(x=1), \mu(x=1)) = p_Q(T(x=0), \mu(x=0))$ along $(T(x), \mu(x))$. It is clear from the

---
[6] Another approach tried recently in order to improve compatibility of lattice results with the HRG is the addition of yet unknown hadron states via the introduction of an exponentially growing hadron mass spectrum [Maj10]. However, this approach has not yet been generalized for finite chemical potential. For this, the unknown hadron contribution would have to be split into a mesonic and a baryonic part introducing another unknown parameter.

continuous structure of the convex combination that the other thermodynamic quantities such as the net particle, entropy or energy density are continuous along the isentropic trajectory as well. A prudent choice for the pressure p_M is the HRG pressure at the intersection of the HRG adiabate with the predicted freeze-out curve [Cle06] which turns out to be $p_M = 0.07$ GeV/fm^3. The corresponding trajectories (T, μ) as well as the resulting EOS can be seen in Fig. 5.14 for both the effective and the HTL QPM.

The result is compatible with the investigations put forward in [Ton03, Non05] but additionally provides very close contact to first-principle lattice results. For these cases, reheating is observed. It is interesting to note that, although the eQPM results are very similar to the HTL QPM results (in fact, so similar that the mixed-phase trajectories are virtually on top of each other) the adiabates are somewhat different which is mainly due to the small differences in the net particle density. As a consequence, the isentropic EOS $e(p)$ are different above p_M with the HTL QPM showing a larger energy density than the eQPM since the adiabate is found at higher chemical potentials (cf. Fig. 5.7).

Note that the EOS does not account for a possible critical point. However, it could be implemented as demonstrated in [Blu05b, Blu06, Blu08c] for the eQPM.

5.6 Equation of state for FAIR

The freeze-out points for the CBM experiment at the heavy-ion synchrotron SIS300 of the planned facility FAIR are at even higher densities than for SPS. For the top energies of 45 AGeV one may assume $\mu_q = 330$ MeV as an approximation for the transition chemical potential [Iva06, Sta10]. The region of negative pressure does not constitute an obstacle as the transition might be constructed at temperatures higher than those where the pressure is negative.

However, the chemical potential $\mu_q = 330$ MeV corresponds to an entropy per particle $s/n_q \approx 3$ for which the isentropes of HRG and either of the two QPM are about another 300 MeV apart. Due to this we have to conclude that a reasonable construction of an EOS for CBM experiments at FAIR, obtained from lattice results using our quasiparticle models and the HRG for the confined phase, is not possible.

5.7 Application in QCD sum rule calculations

A different field of application for the quark-gluon plasma EOS obtained from the HTL QPM constitutes the calculation of gluon condensates for the evaluation of QCD sum rules. This work was performed in collaboration with T. Hilger [Hil10]. According to the consideration in [Mor08] one may relate the gluon condensate at finite temperature to the QCD trace-anomaly

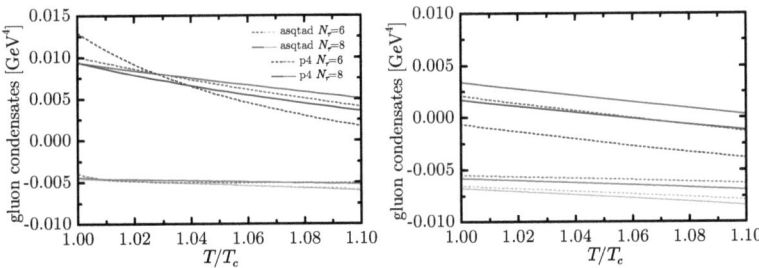

Figure 5.15: Temperature dependence of the gluon condensates (5.2) (upper curves) and (5.3) (lower curves) at $\mu_q = 0$ (left panel) and $\mu_q = 270$ MeV (right panel) for the adjustment of the HTL QPM to the lattice results from [Baz09] (color coding in left panel).

for $N_f = 3$ and the equation of state:

$$\left\langle \frac{\alpha_s}{\pi} G^2 \right\rangle_T = \left\langle \frac{\alpha_s}{\pi} G^2 \right\rangle_0 - \frac{8}{9}(e - 3p), \quad (5.2)$$

$$\left\langle \frac{\alpha_s}{\pi} \left((uG)^2 - \frac{G^2}{4} \right) \right\rangle_T = -\frac{3}{4} \frac{\alpha_s}{\pi}(e + p), \quad (5.3)$$

whereas contributions from light quarks to (5.2) have been omitted in a first step, as we focus on the continuation to finite densities. The quantities $e - 3p$ and $e + p$ at finite chemical potential follow from the HTL QPM – here adjusted to the various lattice results from [Baz09]. Both condensates (5.2) and (5.3) are depicted in Fig. 5.15 in the region near T_c at quark chemical potential $\mu_q = 0$ (left panel) and $\mu_q = 270$ MeV (right panel).

The curves for the condensate (5.2) are flattened with increasing lattice temporal extent N_τ; results of [Mor08] are reproduced by the p4 action for $N_\tau = 6$. On the other hand, the condensate (5.3) seems not to be affected by such a choice. At finite densities, the gluon condensate $\langle \alpha_s G^2/\pi \rangle_T$ drops significantly due to the nonzero chemical potential. The symmetric and traceless gluon condensate $\langle \alpha_s \left((uG)^2 - G^2/4 \right)/\pi \rangle_T$ is much less influenced by density effects.

The observed effects influence sum rules for heavy quark objects, e.g. charmonia such as the J/ψ, as these depend essentially on the considered gluon condensates [Mor08]. Such studies are not part of this thesis. For additional details see [Hil10].

5.8 Wrap-up

The eQPM and the HTL QPM show remarkable similarity in flexibility to describe lattice results at $\mu = 0$ (for different parametrizations) and extrapolated state variables. As a consequence, the eQPM can be considered a good approximation to the HTL QPM and represents a decent tool for the exploratory study put forward in Chapter 6.

Using the HTL QPM as description of the quark-gluon phase an EOS for SPS was obtained by a standard mixed-phase construction to the hadron resonance gas. Due to the large distances of HTL QPM and HRG adiabates a similar construction for the CBM experiment at FAIR is not sensible.

One possible application of the results in the evaluation of QCD sum rules at finite temperature was presented. Not shown explicitly is the verification of model consistency as demonstrated for the eQPM (cf. Section 4.4). As to be expected the results are similar and model consistency is assured.

Lattice results including a charm quark [Bor10b] suggest non-negligible contributions a low as $T \sim 200$ MeV. At $T \sim 0.8$ GeV the effect is estimated to be of about 20 percent. While therefore some changes can be expected from including the charm quark in the analysis, noticeable improvement, especially in the relevant region close to the phase transition, is unlikely.

6 Compact stellar objects

After growing evidence for the quark-gluon substructure of hadrons the question has been asked [Ito70, Bay76, Kei76, Fre78, Fec78] whether massive neutron stars may have a core composed of quarks [Gle97, Gle00, Web99, Web05]. These so-called hybrid stars may be part of the neutron star branch or constitute a separate stable branch of high-density objects – the so-called third family [Ger68, Kam81a, Kam81b, Kam83, Kam85] or twin stars [ScB02]. Also pure quark stars populating another separate branch of stable, spherically symmetric cold objects have been discussed [Fec78, Ana79, Pes00]. All these possibilities depend sensitively on the EOS at high density and the details of the deconfinement transition at low temperature. While at high temperature and zero net baryon density a proper numerical evaluation of the equation of state using lattice methods based on first-principle QCD is accomplished, the knowledge of the equation of state at high baryon density and low temperature is fairly poor.

In the asymptotic region, safe statements on the matter states can be made [SW99, Ris01, Raj01], but the extrapolation to the interesting region of energy densities around 10^{15} g/cm^3 is hampered by serious uncertainties as one expects significant non-perturbative effects. Using low-order perturbative expansions, approximations to the EOS at vanishing temperature have been obtained [Fra01, Fra02, AS02], however, these calculations have little predictive power, as the scale parameter of the theory cannot be fixed without connection to (currently still lacking) experiments.[1]

Due to its self-consistency, the QPM enables us to map lattice QCD results for the thermodynamic state variables, and thus also the EOS, available at vanishing chemical potential, to large μ and also small temperatures (cf. Fig. 4.3). In the farthest case, an EOS for cold stellar matter can therefore be derived from the first-principle lattice calculations. In a way, the QPM is thus able to close a gap by taking the lattice QCD results as (approximate) "experimental" evidence and compare the outcome of the extrapolation with the parametric perturbative results, yielding an approximation for the scale parameter. Analog studies have been performed in, e.g., [SLS99, Pes01b, Pes03, Iva05], however without such intimate contact to advanced lattice QCD results.

After recalling some basic concepts and notations in Sections 6.1 and 6.2 we introduce a

---

[1] Supplementing principles may be employed to adjust the scale parameter. For instance, [Kur09] argues that the quark pressure has to of the order of the pressure of the hadronic phase at the maximum of the latter when scaled with μ^4 in order to allow for a straightforward construction of the phase transition. Our results indicate that this may not be the case.

modified QPM, allowing for a simultaneous treatment of the weakly interacting sector. Using the found EOS we investigate the properties of the resulting pure quark and hybrid stars and derive general arguments concerning their (non)existence.

6.1 Gravitation and general relativity

The fundamental force of gravitation is best described by the theory of general relativity, where energy (including mass energy and momentum) is the source of gravitation by means of a deformation of a 4-dimensional pseudo-Riemann manifold representing spacetime. Particles are not subjected to any fields but rather move freely along geodesics (shortest paths between two points) in the curved space – which do not necessarily have to be straight. For instance, flight routes appear curved on a Mercator projection as the shortest connection on a sphere is a great circle route. In the same way the periodic orbit of a planet around a star is a geodesic.

The deformation of a manifold is quantified by the curvature tensor $R_{\mu\nu\sigma\omega}$ – given in terms of the metric $g_{\mu\nu}$ via Christoffel symbols – which measures the deviation from an Euclidean space and its contractions, the Ricci-Tensor $R_{\mu\nu} := R^{\sigma}{}_{\mu\sigma\nu}$ and the curvature scalar $R := R_{\mu\nu}R^{\mu\nu}$. In order for the curvature tensor, which contains 20 independent components, to describe an actual metric with only 10 metric functions, it has to satisfy 10 Bianchi identities from which, by contraction with the metric tensor, the requirement that the special combination $G_{\mu\nu} := R_{\mu\nu} - \frac{1}{2}Rg_{\mu\nu}$, dubbed Einstein tensor, be divergence-free ($G^{\mu\nu}{}_{;\mu} = 0$, where the semicolon denotes the covariant – curvature-sensitive – derivative) follows.

The Einstein equations

$$G_{\mu\nu} = 8\pi G_N T_{\mu\nu}, \qquad (6.1)$$

where G_N is Newton's gravitational constant, describe how exactly energy, quantified by the energy-momentum tensor $T_{\mu\nu}$, modifies the curvature of spacetime. The integrability condition of the contracted Bianchi identities that $G_{\mu\nu}$ be divergence-free translates to the energy-momentum tensor and gives the continuity equation

$$T^{\mu\nu}{}_{;\mu} = 0, \qquad (6.2)$$

i.e. local energy-momentum conservation, which leads to the geodesic equation (matter moves along geodesics as the shortest paths) as the equation of motion of matter.

The Einstein equations are nonlinear partial differential equations and only few exact solutions exist. For the solution it is necessary to assume at the same time a specific energy-momentum tensor and a metric tensor, which on the one hand is needed to evaluate (6.2) and on the other hand has to satisfy (6.1) where again the energy-momentum tensor enters. The problem per se is under-determined: only six of the ten Einstein equations are independent as the Einstein tensor has to satisfy the four contracted Bianchi identities while ten independent component of $g_{\mu\nu}$ need to be determined. This is due to the invariance of general relativity under coordinate transformations of the metric tensor which can be used to to choose a coordinate system suitable for the particular physical situation.

6.2 The Tolman-Oppenheimer-Volkoff equations

Here we are interested in the mass-radius relation of compact stellar objects which are cold and spherically symmetric. A general relativistic description of a static, spherical compact stellar object is given by the Tolman-Oppenheimer-Volkoff (TOV) equations [Tol34, Tol39, OV39] which follow from the Einstein equations and the Bianchi identities by assuming the energy momentum tensor of an ideal fluid $T^{\mu\nu} = (e+p)u^\mu u^\nu - pg^{\mu\nu}$, rotational symmetry (in $g^{\mu\nu}$) and statics ($u^\mu \sim (u^0, 0)$). They present a system of two coupled ordinary differential equations

$$\frac{dp}{dr} = -G_N \frac{(e+p)(m + 4\pi r^3 p)}{r^2(1 - \frac{2m}{r} G_N)}, \qquad (6.3)$$

$$\frac{dm}{dr} = 4\pi r^2 e, \qquad (6.4)$$

where G_N is the Newtonian gravitational constant and we employ units with $\hbar c = 1$, which by itself is under-determined. Again the equation of state, given as $e = e(p)$, is needed in order to solve the problem.

With a given equation of state the solution is straightforward using standard numerical techniques. Starting with initial conditions $p_c := p(r = 0)$ and $m(r = 0) = 0$ the Eqs. (6.4) are integrated along the radial coordinate r until the pressure falls to zero. This is exactly the boundary of the sphere, so that the definition of the radius R of the object is

$$p(R) := 0. \qquad (6.5)$$

The quantity

$$m(r) = 4\pi \int_0^r dr' \, r'^2 e(p(r')) \qquad (6.6)$$

from the second TOV equation integrates the mass inside the integration limits. Therefore the mass of the object is defined as

$$M := m(R) \qquad (6.7)$$

which enters directly the exterior Schwarzschild solution and determines, e.g. the trajectories of test particles and light

6.3 Including the weak sector

In order to describe quark matter in compact stellar objects it is necessary to take the weakly interacting sector into account. For this, the QPM has to be modified to describe a plasma of gluons, quarks and leptons in equilibrium. As shown in Chapter 5, the results of the effective and the HTL QPM are quite similar. In view of the analytically much more involved structure of the HTL QPM it is most prudent to (at first) use the eQPM as underlying QPM for this extension. In addition, the eQPM allows for the easy inclusion of quasiparticle restmasses.

We consider here the contributions from electrons and muons and assume that the neutrinos $\nu_{e,\mu}$ left the star matter and, therefore, do not participate in the chemical equilibrium reactions. Due to its large mass, adding the tau does not lead to notable changes. In this chapter, the assumption of equal light quark and zero heavy chemical potentials is lifted, giving five independent chemical potentials.

The four relations for (i) charge neutrality

$$\sum C_i n_i^{eQP+l} = 0 \tag{6.8}$$

(with $C_u = 2/3$, $C_d = C_s = -1/3$ and $C_e = C_\mu = -1$), for (ii) β equilibrium

$$\mu_d = \mu_u + \mu_e \tag{6.9}$$

(e.g., from $n \leftrightarrow p^+ + e^- + \bar{\nu}_e$), for (iii) equilibrium due to strangeness changing weak decays

$$\mu_s = \mu_d \tag{6.10}$$

(e.g., from $\Lambda \leftrightarrow p^+ + \pi^-$) and for (iv) μ decay

$$\mu_\mu = \mu_e \tag{6.11}$$

(e.g., from $\mu^- \leftrightarrow e^- + \bar{\nu}_e + \nu_\mu$) map the various chemical potentials on just one independent chemical potential μ via functions $\mu_{u,d,s,e,\mu}(\mu)$ – some of which are only given implicitly. We choose $\mu = \mu_u$ from which μ_d and μ_s follow directly with $\mu_e = \mu_\mu$ being fixed via the electric charge neutrality condition (6.8). The latter constitutes an implicit equation that cannot be analytically solved for μ_e, but has to be determined numerically for any set of (T, μ, G^2).

For brevity, we dub the modified model eQPM+l, referring to the added leptons. Keeping the assumption of non-interacting quasiparticles, the thermodynamic quantities are given as sums of the single quasiparticle quantities

$$\sum_{i=g,q,s} \longrightarrow \sum_{i=g,u,d,s,e,\mu} \tag{6.12}$$

with electron and muon contributions assumed to follow ideal gas expressions. Due to the different chemical potentials, up and down quark contributions to the thermodynamic quantities, e.g. n_i^{eQP+l} used in Eq. (6.8), have to be calculated separately with degeneracy factors $d_u = d_d = 2N_c = d_s$ instead of $d_q = 2N_c N_l$ (cf. Section 2.6) as well as the strange quark contributions

6.4 Extrapolation to nonzero chemical potential

which now also have a nonzero chemical potential.

The thermal fermion masses, which enter the asymptotic quark masses $\tilde{m}_{i,\infty}^2 = m_{i,0}^2 + 2m_{i,0}\hat{M}_i + 2\hat{M}_i^2$ (cf. Eq. (3.6)) in the eQPM dispersion relations $\omega_i^2 = k^2 + \tilde{m}_{i,\infty}^2$, are (cf. Eq. (2.19))

$$M_i^2 = \frac{C_f}{8}\left(T^2 + \frac{\mu_i^2}{\pi^2}\right)G^2. \tag{6.13}$$

The summation over the chemical potentials in the Debye mass (cf. Eq. (2.19)) runs through all three quark flavors. Employing the side conditions (6.9) and (6.10) yields the asymptotic gluon mass (cf. Eq. (3.3))

$$\tilde{m}_{g,\infty}^2 = m_{g,\infty}^2 = \left(\frac{C_b}{6}T^2 + \frac{N_c}{12\pi^2}\left(3\mu^2 + 4\mu\mu_e + 2\mu_e^2\right)\right)G^2. \tag{6.14}$$

At $\mu = 0$, the net particle density of up quarks is zero. Since all other considered particles have negative electric charge, the electric neutrality condition (6.8) requires $\mu_e = 0$. From this it is clear that the quark and gluon contributions at vanishing chemical potential are equivalent for eQPM+l and eQPM. In addition, since lattice results describe only QCD matter, the adjustment of model parameters to the lattice results has to exclude the electron and muon contributions to interaction measure and pressure, leaving an expression identical to the eQPM. As a result, the obtained adjustments of model parameters to lattice calculations in Chapter 4 (cf. Tab. 4.1) are also valid for the eQPM+l. In fact, the contribution of electrons and muons to the interaction measure is very small (cf. right panel of Fig. 6.3) so that the only notable difference would be a marginally modified pressure constant B_c.

6.4 Extrapolation to nonzero chemical potential

Going to non-vanishing chemical potential, the – formally equivalent – flow equation (2.50) contains separate contributions of up, down and strange quarks[2]

$$\sum_{i=g,u,d,s}\frac{\partial s_i}{\partial \tilde{m}_i^2}\frac{\partial \tilde{m}_i^2}{\partial \mu} = \sum_{i=u,d,s}\frac{\partial n_i}{\partial \tilde{m}_i^2}\frac{\partial \tilde{m}_i^2}{\partial T}, \tag{6.15}$$

where the index "∞" is suppressed for brevity. Note that the derivatives of the asymptotic masses are with respect to the up quark chemical potential μ so that $\partial \tilde{m}_i^2/\partial \mu = (\partial \mu_i/\partial \mu)(\partial \tilde{m}_i^2/\partial \mu_i)$. As the thermodynamic quantities, the derivatives of the latter equal the eQPM expressions with $d_{\{u,d,s\}} = 2N_c$ (cf. Section 6.3). For brevity we have omitted the notation $eQP+l$ of the state variables which shall be implied for the remainder of this chapter.

---

[2] The flow equation does not contain the electron and muon contributions where the pressure is not given including a mean field contribution. Therefore, electron and muon pressure are a priori thermodynamic potentials and would therefore immediately vanish from the flow equation due to the theorem of Schwarz.

The derivatives of the asymptotic masses with respect to the temperature, the effective coupling and the respective chemical potential are given in Appendix C. Due to the dependence of gluon, down and strange quark masses on the electron chemical potential, the derivatives of the latter with respect to temperature, (up quark) chemical potential and effective coupling are required. Since μ_e is determined via the electric charge neutrality condition (6.8), which is an implicit equation, the derivatives follow as

$$\left.\frac{\partial \mu_e}{\partial T}\right|_{\mu,G^2} = -\left(\sum_i C_i \left.\frac{\partial n_i}{\partial \mu_e}\right|_{T,\mu,G^2}\right)^{-1} \left(\sum_i C_i \left.\frac{\partial n_i}{\partial T}\right|_{\mu_e,\mu,G^2}\right),$$

$$\left.\frac{\partial \mu_e}{\partial \mu}\right|_{T,G^2} = -\left(\sum_i C_i \left.\frac{\partial n_i}{\partial \mu_e}\right|_{T,\mu,G^2}\right)^{-1} \left(\sum_i C_i \left.\frac{\partial n_i}{\partial \mu}\right|_{\mu_e,T,G^2}\right),$$

$$\left.\frac{\partial \mu_e}{\partial G^2}\right|_{T,\mu} = -\left(\sum_i C_i \left.\frac{\partial n_i}{\partial \mu_e}\right|_{T,\mu,G^2}\right)^{-1} \left(\sum_i C_i \left.\frac{\partial n_i}{\partial G^2}\right|_{\mu_e,\mu,T}\right) \qquad (6.16)$$

with sums running over all plasma constituents $i = u, d, s, e, \mu$ and

$$\left.\frac{\partial n_i}{\partial \mu_e}\right|_{T,\mu,G^2} = \left.\frac{\partial \mu_i}{\partial \mu_e}\right|_\mu \left(\left.\frac{\partial n_i}{\partial \mu_i}\right|_{m_i^2} + \frac{\partial n_i}{\partial m_i^2} \left.\frac{\partial m_i^2}{\partial \mu_i}\right|_{G^2}\right),$$

$$\left.\frac{\partial n_i}{\partial \mu}\right|_{T,\mu_e,G^2} = \left.\frac{\partial \mu_i}{\partial \mu}\right|_{\mu_e} \left(\left.\frac{\partial n_i}{\partial \mu_i}\right|_{m_i^2} + \frac{\partial n_i}{\partial m_i^2} \left.\frac{\partial m_i^2}{\partial \mu_i}\right|_{G^2}\right),$$

$$\left.\frac{\partial n_i}{\partial T}\right|_{\mu,\mu_e,G^2} = \left.\frac{\partial n_i}{\partial T}\right|_{m_i^2} + \frac{\partial n_i}{\partial m_i^2} \left.\frac{\partial m_i^2}{\partial T}\right|_{G^2},$$

$$\left.\frac{\partial n_i}{\partial G^2}\right|_{\mu,\mu_e,T} = \frac{\partial n_i}{\partial m_i^2} \frac{\partial m_i^2}{\partial G^2} \qquad (6.17)$$

for the quarks and gluons, where, with the indicator function $\mathbf{1}$, $\partial \mu_i/\partial \mu_e|_\mu = \mathbf{1}_{\{u,d\}}(i)$ and $\partial \mu_i/\partial \mu|_{\mu_e} = \mathbf{1}_{\{u,d,s\}}(i)$, i.e. no complications due to the dependence of μ_e on μ (or vice versa) arise since the respective other chemical potential is kept constant in the derivatives.

The same is valid for the mass derivatives which, in contrast to the flow equation, are with respect to the contained chemical potential μ_i and consequently are equivalent the standard eQPM mass derivatives. Thus, using the eQPM mass derivatives to obtain the derivatives of the electron chemical potential, the eQPM+l mass derivatives are found.

The characteristic curves of the eQPM+l flow equation (6.15) are shown in Fig. 6.1. In comparison to the eQPM, the characteristic curves are shifted slightly to towards lower chemical potential. However, as for the former the characteristic curves of the adjustment to [Bor10b] lattice results reach lower chemical potentials due to lower T_c than if adjusting to [Baz09] lattice results.

Figures 6.2 through 6.8 show the thermodynamic quantities of the eQPM+l as results of the extrapolation procedure using the lattice results from [Bor10b] as boundary values within

6.4 Extrapolation to nonzero chemical potential

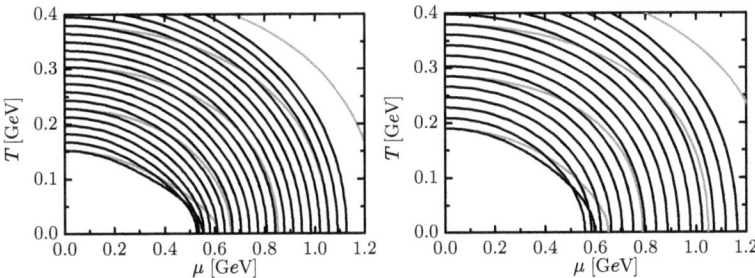

Figure 6.1: Some characteristic curves $(T(x), \mu(x))$ of the flow equation (2.56), emerging from the temperature interval $T_0 \epsilon [T_c, 450\,\text{MeV}]$, are shown for the eQPM+l quasiparticle model (in black). The left curves are for [Bor10b] lattice results with $T_c = 152$ MeV, the right ones for [Baz09] results using the p4 action with $N_\tau = 8$ and $T_c = 190$ MeV. The grey curves indicate characteristics of the eQPM (cf. Fig. 4.3).

the T-μ-plane as functions of the temperature at several values of the chemical potential and additionally as contour plots. In order to separate the contributions of QCD and weak sector and to compare the former to the previous eQPM results (grey curves), the quark-gluon contributions are shown separately as thin curves.

As for the eQPM, almost all state variables (i.e. pressure, interaction measure, entropy density, net particle density and energy density) increase with growing chemical potential for a constant temperature, regardless if considering also electrons and muons or not. The behavior of mean field pressure B and effective coupling G^2 is also unchanged. While the contributions from the weak sector to pressure, energy density and entropy density are substantial, they almost cancel each other in the interaction measure (Fig. 6.3) and the net particle density (Fig. 6.5) which therefore resemble the eQPM result.

From the contour plots we see that the extrapolation procedure yields no artifacts and the information obtained at vanishing chemical potential is transported in a thermodynamically consistent way to $T = 0$ at large chemical potential. As for the eQPM (cf. Section 4.7), the interaction measure (Fig. 6.3) and the pure mean field pressure $B - B_0$ (Fig. 6.7) stay almost constant along the path of the characteristic curves; most notably for $T_0 > 1.5 T_c$.

The lines of constant pressure and the lines of constant energy density still resemble each other (cf. Fig. 6.2 and 6.6), although not as closely as for the eQPM. From Figs. 6.4 and 6.5 it is visible that the limits $s(T \to 0) = 0$ and $n_q(\mu \to 0) = 0$ are obeyed.

A direct comparison of the eQPM+l to the effective and HTL QPM results in the contour plots is not sensible, as the latter two do not account for the weak degrees of freedom. For instance the substantial contributions of electrons and muons at $\mu = 0$ almost vanish along the characteristic curves, leading to lines of constant pressure decreasing faster in temperature with increasing chemical potential (cf. Figs. 4.4 and 6.2).

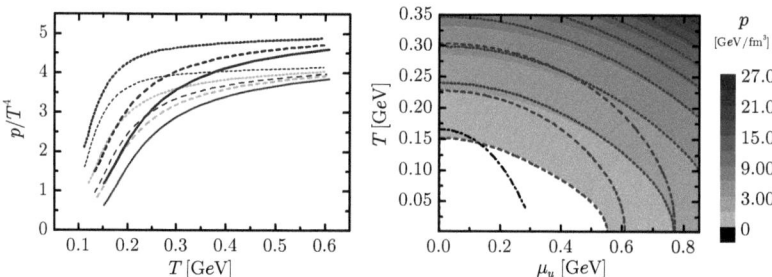

Figure 6.2: Left panel: The scaled overall pressure p/T^4 of the eQPM+l adjusted to [Bor10b] lattice results (dark thick curves) and the quark-gluon contribution (dark thin curves) are shown as functions of the temperature T for chemical potential $\mu = 0$ (solid curve), 0.2 GeV (dashed curve) and 0.3 GeV (dotted curve). For reference, the eQPM results, which at $\mu = 0$ match the quark-gluon contribution of the eQPM+l, are shown as well (grey curves, cf. Fig. 4.4). Right panel: Contour plot of the pressure p. The characteristic curves emerging from $T_0/T_c = (1.0, 1.5, 2.0)$ are depicted as dashed curves. The dotted curves indicate the lines of constant quark-gluon contribution. The chemical freeze-out curve (dotted curve) from the statistical model in [Cle06] is plotted, too.

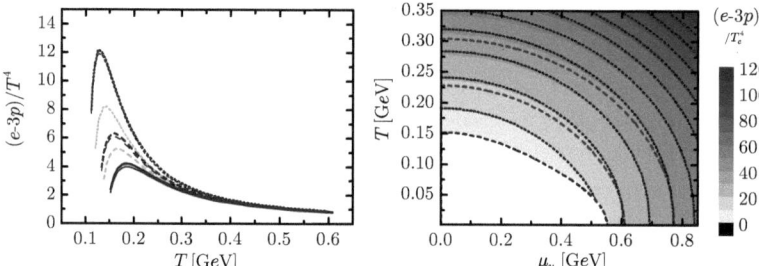

Figure 6.3: As Fig. 6.2 but for the scaled overall interaction measure $(e - 3p)/T^4$. For the eQPM results (grey curves in the left panel) cf. Fig. 4.5.

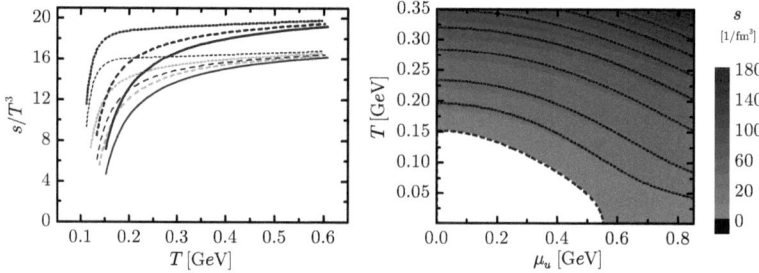

Figure 6.4: As Fig. 6.2 but for the scaled overall entropy density s/T^3. For the eQPM results (grey curves in the left panel) cf. Fig. 4.6.

6.4 Extrapolation to nonzero chemical potential

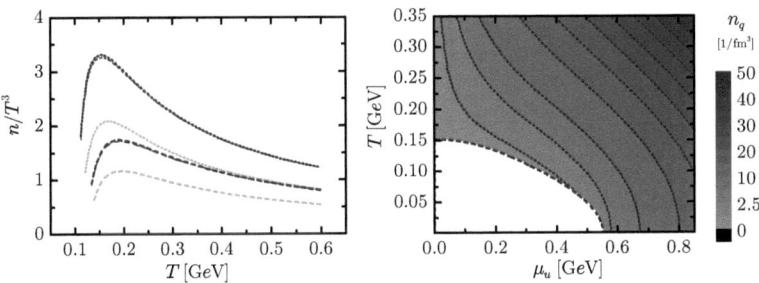

Figure 6.5: As Fig. 6.2 but for the scaled overall net particle density n/T^3. For the eQPM results (grey curves in the left panel) cf. Fig. 4.7. The dotted curves indicate the lines of constant net quark density.

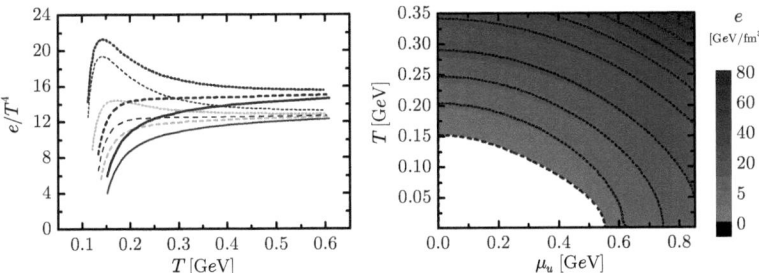

Figure 6.6: As Fig. 6.2 but for the scaled overall energy density e/T^4. For the eQPM results (grey curves in the left panel) cf. Fig. 4.8.

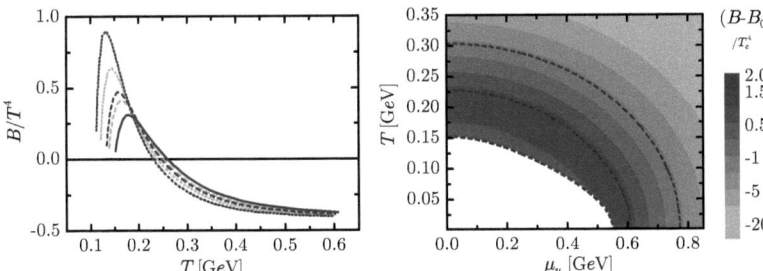

Figure 6.7: As Fig. 6.2 but for the scaled mean field pressure B/T^4. As there is no contribution from the leptons, overall pressure and quark-gluon contribution are equal. For the eQPM results (grey curves in the left panel) cf. Fig. 4.9.

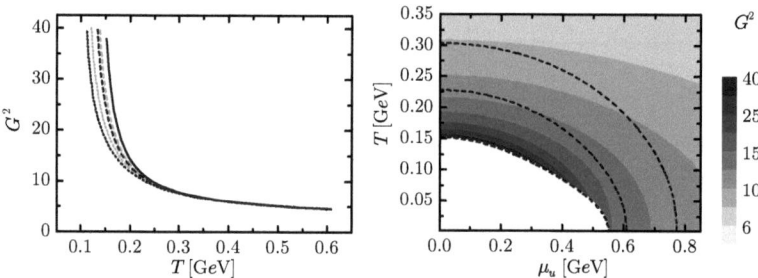

Figure 6.8: As Fig. 6.2 but for the effective coupling G^2. For the eQPM results (grey curves in the left panel) cf. Fig. 4.10.

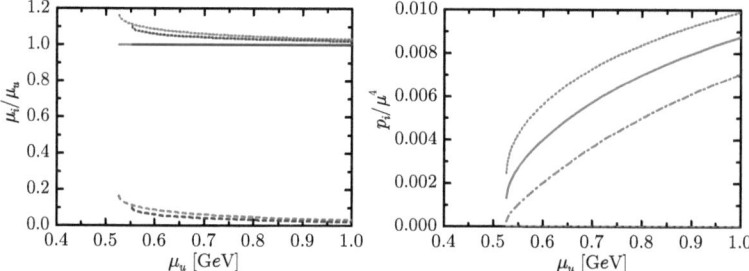

Figure 6.9: Left panel: The chemical potentials of the plasma constituents (solid curves: up quarks, dashed curves: electrons and muons, dotted curves: down and strange quarks) scaled with the up quark chemical potential $\mu = \mu_u$. Green curves are for the adjustment to [Bor10b] lattice results while blue curves are for the adjustment to lattice results from [Baz09] using the p4 action with $N_\tau = 8$. Right panel: Scaled partial pressures of up (solid curve), down (dotted curve) and strange (dash-dotted curve) quarks as well as electrons and muons (dash-dotted curves, identical with the μ_u-axis on the chosen scale) for the adjustment of the eQP+l to [Bor10b] lattice results.

6.5 State variables at zero temperature

At $T = 0$, the contributions from the weak sector are suppressed due to small electron chemical potential and lower degeneracy factors. Even at the low chemical potentials, where μ_e/μ has its maximum ($\mu_e/\mu = 0.16$ for the adjustment to [Bor10b] lattice results and 0.1 for the adjustment to lattice results from [Baz09] using the p4 action with $N_\tau = 8$, cf. Fig. 6.9), the contributions of electrons and muons to the thermodynamic quantities are only of about 2 to 3 percent. For example, the partial pressures of quarks and leptons are plotted in the right panel of Fig. 6.9.

The electron and muon contributions are not visible and it is therefore not necessary to distinguish between pure QGP and QGP+weak sector results for the thermodynamic quantities at $T = 0$. This is in line with arguments put forward in [SW99]. Results for the energy and net particle density are similar in that the down quark contributions turn out to be the largest.

6.5 State variables at zero temperature

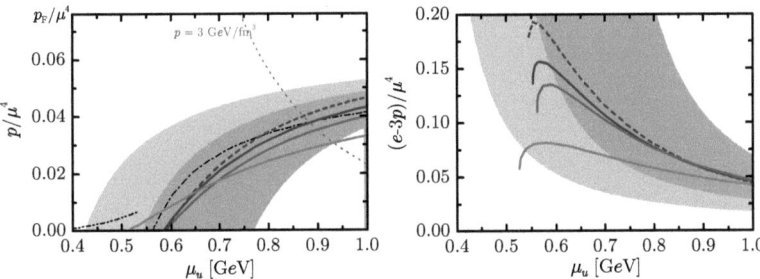

Figure 6.10: Scaled pressure p/μ^4 (left panel) and scaled interaction measure $(e-3p)/\mu^4$ (right panel) of the eQPM+l as functions of $\mu = \mu_u$ at $T = 0$ for several adjustments to lattice results in comparison to results from [AS02] (dash-dotted curve), [SLS99] (dash-double-dotted curve) and [Fra01, Fra02] (grey bands limited by $\bar{\mu}/\mu = 1$, 1.5 and 2 from dark to light). Green curves are for the adjustment to [Bor10b] lattice results while blue/red curves are for the adjustment to lattice results from [Baz09] using the p4/asqtad action, respectively, where dashed/solid lines denote $N_\tau = 6/8$. In addition the line of constant pressure $p = 3$ GeV/fm^3 as approximate upper limit of pure quark star central pressures is given in the left panel (cf. Section 6.8).

Due to smaller chemical potential, the up quark contributions are somewhat smaller while, due to the large restmass, the strange quark contributions are even more suppressed. In contrast, for the energy density, the up and strange quark contributions are very close (not shown). It happens that the effective coupling G^2 can also be parametrized at vanishing temperature using Eq. (2.31) but with $(T - T_s)/\lambda \to (\mu - \mu_s)/\lambda_\mu$. The parameters are listed in Tab. 6.1.

Figs. 6.10 and 6.11 show the predictions from the eQPM+l adjusted to [Baz09] and [Bor10b] lattice results for the thermodynamic quantities at vanishing temperature as applicable for compact stellar objects. For the scaled pressure, the results of the various adjustments are very close, especially the results from [Baz09] which share a common chemical potential of about 0.59 GeV at vanishing pressure. The pressure from the adjustment to [Bor10b] is approaching zero for small chemical potential, however, does not reach zero and rather stops – with for the result from the last non-crossing characteristic – at chemical potential $\mu_u = 0.52$ GeV and pressure $p = 8.3$ MeV/fm^3. The slope of the pressure is larger as well for the results for the adjustment to [Baz09] lattice data, however, the smaller initial chemical potential and the lesser slope for the results for the adjustment to [Bor10b] somewhat cancel each other, yielding a result within the range of the other predictions.

It is most notable in the scaled interaction measure that the results for the thermodynamic quantities at $T = 0$ are, by means of scales and shifts, similar to the results at $\mu = 0$. As for the interaction measure at vanishing chemical potential (Fig. 4.1) the p4 $N_\tau = 6$ result displays the highest peak, with the $N_\tau = 8$ and asqtad case following and the stout action result having the smallest peak at somewhat smaller temperature/chemical potential.

Our results can be compared with perturbative calculations at vanishing temperature in [AS02, Fra01, Fra02]. In [AS02] the pressure of cold quark matter is calculated in hard-dense-

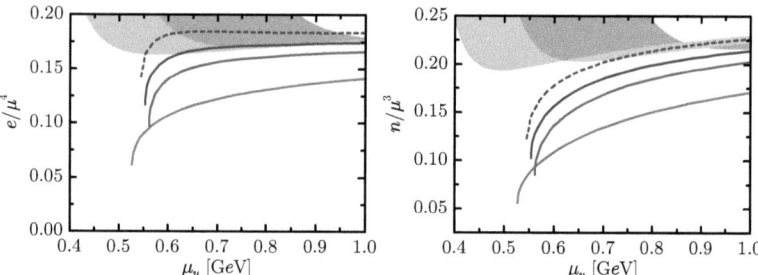

Figure 6.11: Scaled energy density e/μ^4 (left panel) and scaled net particle density n/μ^3 (right panel) of the eQPM+l as functions of $\mu = \mu_u$ at $T = 0$ for several adjustments to lattice results. Green curves are for the adjustment to [Bor10b] lattice results while blue/red curves are for the adjustment to lattice results from [Baz09] using the p4/asqtad action, respectively, where dashed/solid lines depict $N_\tau = 6/8$.

loop perturbation theory. The resulting pressure for 3 flavors with equal chemical potential and the choice of the renormalization scale $\bar{\mu} = \mu$ [AS02] is shown in the left panel of Fig. 6.10. Also shown are the results from the weak-coupling expansion to second order [Fra01, Fra02]. Here the value of $\bar{\mu}$ is varied from μ to 2μ. For reference, also a comparison of the pressure with NJL model results [SLS99] is depicted. The grey dotted curve is $p = 3$ GeV/fm^3; the region $0 < p < 3$ GeV/fm^3 is relevant for quark stars, as turns out by integrating the TOV equations (see Section 6.8).

The results for the pressure are well within the range of the perturbative predictions for relevant pressures $p < 3$ GeV/fm^3. The same is true for the interaction measure, where, however, the increase at small chemical potentials, extrapolated from the non-perturbative phase transition region close to T_c at $\mu = 0$, is missing in the perturbative predictions. This is also part of the reason, why the agreement with the perturbative predictions is less optimal for the energy and net quark density. At $T = 0$ the energy density

$$e|_{T=0} = \mu^2 \frac{\partial}{\partial \mu} \frac{p}{\mu} \qquad (6.18)$$

depends on the slope of the pressure scaled with the chemical potential rather than the absolute values of the pressure. Due to the large slopes of the perturbative predictions for the pressure, the region predicted for the energy density indicate an area matched only by the results for adjustments to p4 action lattice data. Due to $n/\mu^3 = (e+p)/\mu^4$ (at $T = 0$) this also translates to the net particle density.

As for the interaction measure, the slope of the quantities derived from the pressure at the smaller chemical potentials is exactly opposite to the QPM prediction, as the non-perturbative effects in the phase transition region are not contained within the purely perturbative calculations. If, however, the slope of those quantities is not predicted in a compatible manner, a quantitative comparison is not reasonable. Comparing the energy density and net quark density from eQPM+l and [Fra01, Fra02] qualitatively outside of the non-perturbative region, we find a general agreement e.g. in slopes and asymptotic behavior.

Not shown is the comparison of eQPM+l with eQPM results. As a general rule, the eQPM results are two third to one half of the eQPM+l results, mostly due to the additional strange quark and the enlarged down quark contribution (cf. Fig. 6.9 for the partial pressures). A verification of model consistency as in Section 4.4 can also be performed for the eQPM+l model and yields similar results.

6.6 Equation of state

The resulting eQPM+l equation of state at $T = 0$ in the form $e(p)$, needed for the integration of the TOV equations below, is exhibited in Fig. 6.12 (left panel for a comparison of the several adjustments) together with two fits by $e = v_s^{-2} p + e_0$ for the equations of state adjusted to p4 $N_\tau = 8$ and stout action (right panel) as well as the EOS predictions from [AS02] and [Fra01, Fra02][3]. As for the pressure and despite the differences in the energy density, the results are well within the range of perturbative predictions. Due to the similar pressure slopes of the adjustments to [Baz09] lattice results, the EOS of the latter are very similar with differences mainly in the vacuum energy density $e_0 := e(p = 0)$. Asymptotically, also the EOS for the adjustment to [Bor10b] lattice results approaches the other results, however, this region is not relevant for compact stellar objects anymore.

The fit parameters for all actions and temporal lattice extents considered here are listed in Tab. 6.2. Both, the vacuum energy density $e_0 = e(p = 0)$ and the velocity of sound parameter v_s^2 ($= \partial p/\partial e$ for the linear fit) are in narrow intervals for the [Baz09] lattice QCD input data. The largest deviations of the three equations of state, found at vanishing pressure, are about 10%. While the vacuum energy density varies within $(375 \text{ MeV})^4$ to $(395 \text{ MeV})^4$, v_s^{-2} is within 3.4 to 3.7. In comparison, the inverse square of the velocity of sound for the [Bor10b] lattice QCD input data is slightly larger while the vacuum energy density of the fit is somewhat smaller with about $(320 \text{ MeV})^4$.

As for the state variables in the previous section, the contributions of electron and muons to the EOS are tiny. Also, thermal effects are found to be small, i.e. up to $T = 50$ MeV the equation of state $e(p)$ does not change significantly.

Fig. 6.12 clearly evidences that the previous foundation for discussing quark stars seemed not to be on safe grounds as the proposed model equations of state were too different unless further constraints (e.g. the compatibility with a hadronic model equation of state as required in

---

[3] Inspection of p/μ as a function of μ (not displayed) explains the broad range of values for $e(p \to 0)$ in [Fra01, Fra02]: the slope of p/μ as a function of μ changes drastically with the chosen scale $\bar{\mu}$ for small pressures. For $\bar{\mu} = 1.5\mu$ the equation of state in [Fra01, Fra02] in the form $e(p)$ coincides with the results of [AS02].

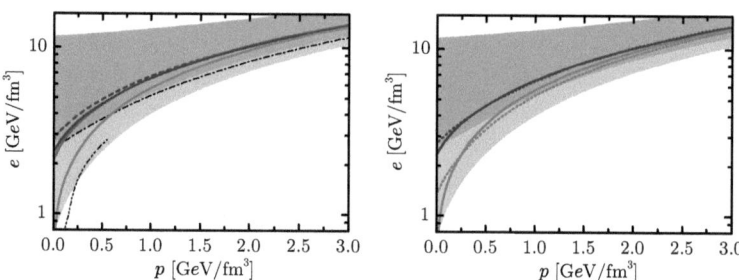

Figure 6.12: Right panel: Equation of state $e(p)$ of the eQPM+l at $T = 0$ for several adjustments to lattice results in comparison to results from [AS02] (dash-dotted curve), [SLS99] (dash-double-dotted curve) and [Fra01, Fra02] (grey bands limited by $\bar{\mu}/\mu = 1$, 1.5 and 2 from dark to light). Green curves are for the adjustment to [Bor10b] lattice results while blue/red curves are for the adjustment to lattice results from [Baz09] using the p4/asqtad action, respectively, where dashed/solid lines denote $N_\tau = 6/8$. Left panel: Same as right panel with some omissions. In addition, fits of two of the EOS by $e = v_s^{-2} p + e_0$ (dotted curves) are shown. For parameters, see Tab. 6.2.

action	N_τ	μ_s [MeV]	λ_μ [MeV]
p4	6	222	164
p4	8	131	227
asqtad	8	−26	369
stout	∞ (est.)	−359	517

Table 6.1: Parameters of Eq. (2.31) with $(T - T_s)/\lambda \to (\mu - \mu_s)/\lambda_\mu$ following from the solution of the flow equation Eq. (6.15) for the adjustments in Tab. 4.1. The fits apply in the range $\mu = 0.6...1.2$ GeV.

action	N_τ	v_s^{-2}	$e_0^{1/4}$ [MeV]
p4	6	3.4	395
p4	8	3.7	381
asqtad	8	3.6	375
stout	∞ (est.)	3.9	320

Table 6.2: Parameters of linear fits $e = v_s^{-2} p + e_0$ to our equation of state with $G^2(\mu)$ determined by the eQPM+l flow equation (6.15). The leptonic contributions are included.

[Fra02]) are imposed. Given the intimate contact of our approach to first-principle evaluations of QCD, we hope to have a more reliable foundation. Of course, this hope is related to the assumption that the extrapolation to nonzero chemical potential is sufficiently smooth. The qualitative agreement of the eQPM with Taylor expansion coefficients for the μ dependence (cf. Section 4.5) as well as the application of our model at imaginary chemical potential [Blu08a] (and not only small values thereof) give us some confidence in our approach.

Finally, let us report on the importance of the side conditions. If one assumes one common chemical potential for all quarks μ and includes leptons ($\mu_e = \mu_\mu$) via a electric neutrality condition $\mu_e = \mu_e(\mu)$, the results of the equation of state differ from the eQPM+l with the side conditions (6.8-6.11) properly invoked on a 10% level.

6.7 Analytic investigation of the TOV equations

Before calculating the properties of pure quark stars for the EOS found in the previous section, let us provide some general arguments on the influence of the EOS. To this end, we start from the TOV equations (6.4) and assume that the special parametrization $e = v_s^{-2} p + e_0$ of the equation of state – which we now know to be reasonable – is supposed to hold.

We find a strong dependence on the actual value of e_0 which determines the pressure gradient in the dimensionless combination $G_N e_0^{1/2}$ (which is of the order of 10^{-39} for the case at hand), which can be seen in writing the TOV equations as

$$\frac{\partial \bar{p}}{\partial \bar{r}} = -\frac{([1+v_s^{-2}]\bar{p}+1)(\bar{m}+4\pi\bar{r}^3\bar{p})}{\bar{r}^2\left(1-\frac{2\bar{m}}{\bar{r}}\right)},$$
$$\frac{\partial \bar{m}}{\partial \bar{r}} = 4\pi\bar{r}^2\left(v_s^{-2}\bar{p}+1\right), \qquad (6.19)$$

with the scaled quantities $p = \bar{p}e_0$, $r = \bar{r}(G_N e_0)^{-1/2}$, $m = \bar{m}(G_N e_0)^{-1/2}G_N^{-1}$. The scaled TOV equations depend only on v_s^{-2}. The solutions for the relevant values of $v_s^{-2} = 3$ and 4 are exhibited in the right panel of Fig. 6.13. With the given scaling, m and r, as well as the surface radius R and mass M (cf. Eqs. (6.5) and (6.7)), shrink with increasing value of $e_0^{1/2}$, while the dependence on v_s^{-2} is moderate within the interval covering the values of Tab. 6.2. Thus the vacuum energy density e_0 is indeed the decisive quantity determining the sizes and the masses of compact stellar objects obeying our special parametrization. To be specific, for $v_s^{-2} = 3 \pm 1$, the scaled maximum mass is 0.004 ± 0.001.

For any given $r \epsilon [0, R]$ of a star the therein contained mass $m(r)$ (cf. Eq. (6.6)) can be separated into the contributions [MTW73]

$$m(r) = m_0(r) + U(r) + \Omega(r), \qquad (6.20)$$

where

$$m_0(r) := 4\pi \int_0^r \frac{r'^2 \mathrm{d}r'}{\sqrt{1 - 2G_N m(r')/r'}} \sum m_{0,i} n_i(p(r')) \qquad (6.21)$$

is the sum over the restmasses of all contained particles (i.e. if they were all infinitely far apart),

$$U(r) := 4\pi \int_0^r \frac{r'^2 \mathrm{d}r'}{\sqrt{1 - 2G_N m(r')/r'}} \left(e(p(r')) - \sum m_{0,i} n_i(p(r')) \right) \qquad (6.22)$$

the internal energy due to the compression of the particles into the star and Ω the energy gain due to the latter (i.e. the gravitational binding energy) which due to Eq. (6.6) results as

$$\Omega(r) := 4\pi \int_0^r \mathrm{d}r\, r'^2 e(p(r')) \left(1 - \left(1 - \frac{2G_N m(r')}{r'}\right)^{-\frac{1}{2}} \right). \qquad (6.23)$$

We adopt the definition of the binding energy of the contained matter at given r from [Lat00]

$$\begin{aligned} BE(r) &:= m_0(r) - m(r) \\ &= -\Omega(r) - U(r) \qquad (6.24) \\ &= 4\pi \int_0^r \mathrm{d}r\, r^2 \left(\frac{m_{Ba} n(p(r'))}{3\sqrt{1 - 2G_N m(r')/r'}} - e(p(r')) \right), \qquad (6.25) \end{aligned}$$

where bound states are characterized by positive values. For simplicity, we assume that – if applying the calculation to pure quark stars – at infinite distance each set of three quarks forms one baryon with restmass $m_{Ba} = m_{n,p} = 940$ MeV.[4] From Eq. (6.25) it is clear that, in order to calculate the binding energy, not only the EOS $e(p)$ but also the relation $n(p)$ is important.

In order to derive general arguments for the binding energy of compact stellar objects, we hold on to our parametrization of the EOS $e = v_s^{-2} p + e_0$ and search for the corresponding relation $n(p)$. To this end, we define the chemical potential μ_0 at vanishing pressure

$$p(\mu =: \mu_0) = 0. \qquad (6.26)$$

Solving the partial differential equation $\mu^2 (p/\mu)' = v_s^{-2} p + e_0$ (cf. Eq. (6.18)) using Eq. (6.26) as boundary condition yields an expression for the pressure as function of μ:

$$p = e_0 \frac{v_s^2}{1 + v_s^2} \left(\left(\frac{\mu}{\mu_0} \right)^{(1+v_s^2)/v_s^2} - 1 \right). \qquad (6.27)$$

---

[4]In the literature [Lat00], there is some discussion whether to use the proton/neutron mass or rather the iron mass divided by 56 which results in $m_{Ba} = 950$ MeV. This has been discussed in detail, e.g. in [Han04]. For our considerations, this has no noticeable effect.

6.8 Pure quark stars

The net particle density is then found by differentiating p with respect to the chemical potential

$$n = \frac{e_0}{\mu_0} \left(\frac{p}{e_0} \frac{1+v_s^2}{v_s^2} + 1 \right)^{1/(1+v_s^2)}. \tag{6.28}$$

This agrees with general thermodynamics, giving at $T=0$ and μ_0, where entropy density and pressure vanish, $n(\mu_0) = e_0/\mu_0$. It is indeed notable that, for the special case of a linear relation $e(p)$ and given pressure root μ_0, there is only one unique $p(\mu)$ and therefore only one distinct relation $n(p)$. In the spirit of the previously introduced scaled variables, we define the scaled net particle density $\bar{n} = \mu_0 n/e_0$ which also depends only on the inverse square of the velocity of sound v_s^2, while the dependence on e_0 is scaled out.

Rewriting the expression for the binding energy (6.25) in terms of the scaled quantities allows to scale the vacuum energy density e_0 out of the binding energy, too. We define $\overline{BE} := BE (G_N^3 e_0)^{1/2}$ and obtain

$$\overline{BE} = 4\pi \int d\bar{r} \, \bar{r}^2 \left(\frac{m_{\text{Ba}}}{\mu_0} \frac{(\bar{p}(1+v_s^2)/v_s^2 + 1)^{1/(1+v_s^2)}}{3\sqrt{1-2\bar{m}/\bar{r}}} - v_s^{-2}\bar{p} - 1 \right) \tag{6.29}$$

which only depends on v_s^{-2} and μ_0 but not e_0.

For general statements about star binding, only the sign of the binding energy is relevant. Therefore, it suffices to investigate the scaled binding energy $\overline{BE}$. As a consequence, for the special linear parametrization of the EOS – which for many cases may be a good approximation – the argument whether a star composed of matter obeying the latter is bound, can be determined solely by the slope of the EOS, i.e. the inverse velocity of sound v_s^{-2}, and the pressure root μ_0.

The color coding in the right panel of Fig. 6.13, where the scaled mass-radius relationship is shown for two relevant values of v_s^{-2}, indicates which parts of the curves represent bound objects. For rather small $\mu_0 = m_{\text{Ba}}/3 \approx 0.31$ GeV still all configurations are bound (blue curves), while for slightly higher $\mu_0 = 1.1 m_{\text{Ba}}/3$ the lightest configurations already turn out to be unbound. If μ_0 is further increased to $1.2 m_{\text{Ba}}/3 \approx 0.38$ GeV only the heaviest configurations survive if $v_s^{-2} = 3$ while, if $v_s^{-2} = 4$, no bound configurations exist anymore. In contrast, the values of μ_0 found in Section 6.5 for our EOS were 0.59 GeV (adjustment to [Baz09] lattice results) and 0.52 GeV (adjustment to [Bor10b] lattice results). Therefore, from these general arguments, we expect the pure quark stars from our EOS to be unbound[5].

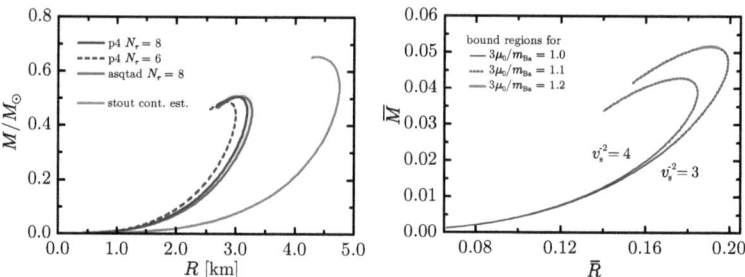

Figure 6.13: Left panel: Solutions of the TOV equations (6.4) using the equation of state of the eQPM+l at $T = 0$ adjusted to several lattice results at $\mu = 0$. Green curves are for the adjustment to lattice results from [Bor10b] while blue/red curves are for the adjustment to lattice results from [Baz09] using the p4/asqtad action, respectively, where dashed/solid lines denote $N_\tau = 6/8$. Right panel: Scaled mass $\bar{M} = \bar{m}(p = 0)$ shown as a function of scaled radius $\bar{R} = \bar{r}(p = 0)$ for several values of v_s^{-2} as solution of the scaled TOV equations (6.19). The color coding indicates the regions of the curves which represent bound objects: for $\mu_0 = m_{Ba}/3$ all configurations are bound while for $\mu_0 = 1.1 m_{Ba}/3$ only the red dotted part of the curves represent bound configurations. If $\mu_0 = 1.2 m_{Ba}/3$ bound configurations are only found in the small green region on the curve for $v_s^{-2} = 3$.

6.8 Pure quark stars

Having prepared the essential inputs and arguments in the preceding chapters, this section now considers the specific case of such compact stellar objects made entirely of eQPM+l matter at $T = 0$, i.e. quarks, electrons and muons. Since the contributions from the latter two have been found to be very small (cf. Secs. 6.5 and 6.6) such objects are dubbed pure quark stars.

We investigate the solutions of the TOV equations (6.4) using the exact EOS derived by solving the flow equation (6.15) an displayed in Fig. 6.12. The pure quark phase holds from the core with $p(r = 0) = p_c$ up to the surface, where $p(r =: R) = 0$. As for the adjustment to the lattice results from [Bor10b] the pressure at $T = 0$ does not reach $p = 0$, the surface pressure was adjusted to coincide with the minimal pressure $p(r =: R) = 8.3$ MeV/fm^3 in this case. This is low enough to expect no deviations larger than one percent for quark stars with, in this case, central pressures of $p_c = 1.42$ GeV/fm^3 at maximum stars mass. For the adjustment to p4 $N_\tau = 8$ lattice results we find the central pressure at maximum star mass to be $p_c = 3.15$ GeV/fm^3.

The resulting mass-radius graphs are shown in the left panel of Fig. 6.13 for all four cases. Due to the large values of the vacuum energy density, the stars are rather small (maximum radii of 3 to 5 km depending on the adjustment) and light (top masses of about 0.5 to 0.65

---

[5]One may, in fact, provide values of v_s^{-2} as well as regions of the mass-radius diagram, where compact stellar objects would be bound despite such large μ_0, however, they turn out to populate regions very close the the black hole limit. The EOS describing stars in these regions is far off the one considered here.

6.8 Pure quark stars

times the mass of the sun $M_\odot)^6$. If such objects were to exist, their bulk characteristics would be quite different from canonical neutron stars with masses concentrated at 1.4 $M_\odot$ and radii of 15 km and larger. Therefore, the pure quark stars from our analysis cannot serve as candidates of twin stars discussed in [ScB02].

With the derived scaling law, equations of state with significantly smaller values of e_0 than deduced in our analysis of the lattice QCD results combined with the employed quasiparticle model, would allow for significantly larger masses and radii. The extent of the scaling law can, for instance, be estimated from the results obtained using the adjustment to the lattice results from [Bor10b]: the rather small difference in e_0 to the other adjustments causes a significant change in the maximum sizes of the resulting pure quark stars even though the also enlarged v_s^{-2} favors smaller radii (cf. Tab. 6.2 and Fig. 6.13).

As expected, the quark stars also turn out to be unbound. While the gravitational binding Ω is of the order of minus 10 percent of the mass parameter M the internal energy U turns out to be about 50 percent resulting in largely unbound objects with negative binding energy BE of about 40 percent (cf. Eq. (6.25)) of their mass parameter M in comparison to standard neutron stars with maximum positive binding energies of about 18 percent of their surface mass[7]. For more information cf. [Sch10].

In addition, the influence of color super conductivity was investigated by adding a pressure term which accounts for the condensation energy of Cooper pairs in the Color-Flavor-Locked (CFL) phase [Alf08]

$$p_{\text{CFL}} = \frac{3\gamma^2}{\pi^2}\mu^2 \tag{6.30}$$

with the gap parameter $\gamma = 0.1$ GeV assumed to be constant which suffices for an estimate of the magnitude of the effects of superconductivity (cf. also [Kur09]). Due to the overall enhancement of the pressure, the resulting scaled pressure p/μ^4 for the model adjusted to the lattice results for the p4 action with $N_\tau = 8$ [Baz09] as function of the chemical potential is missing the root, as it is the case if adjusting to [Bor10b] lattice data (without the CFL phase). The approximate change of the pressure root μ_0 (if extrapolating the pressure to $p = 0$) is from 0.59 GeV to about 0.55 GeV. Even if approximating the same difference onto μ_0 for the [Bor10b] results would yield about 0.48 GeV for the pressure root, still far away from the (at most) allowed 0.38 GeV for bound objects, as derived in Section 6.7, and thus not change the outcome.

---

[6] If considering rapidly rotating pure quark stars instead of the static configuration under investigation here, the maximum masses are enlarged by approximately 65% and the radii by a similar amount [Ans03] at the shedding limit. The stars exhibit a disc like shape with sharp edge.

[7] In [Gle00] it is argued that even though a star (as considered here) is unbound with respect to the simultaneous removal of all constituents to infinity it may still be bound with respect to the removal of single or groups of the latter. Another investigation in [Bom04] claims that the transition from a pure quark star to a hybrid or neutron star may occur on timescales of the order of the age of the universe. Therefore, one may hope that such pure quark stars still exist as meta-stable compact stellar objects. However, no convincing genesis leading to such meta-stable objects is known.

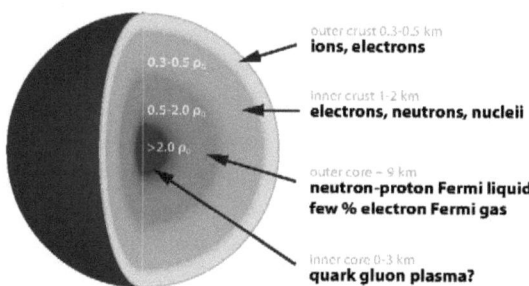

Figure 6.14: Structure of a neutron star. Baryon densities are given in terms of the nuclear saturation density. Created after [Hae07].

6.9 Hybrid approaches

While pure quark stars would have a very distinct border due to the nonzero net particle and energy densities at zero pressure for the quark EOS, the net particle and energy density of a conventional neutron star vanish, as property of the hadronic EOS, with the pressure when approaching the surface, thus creating a rather dilute star boundary. As opposed to the pure quark stars, such ordinary neutron stars are bound objects.

However, even for the latter the forces in the core are strongly repulsive and much larger than the gravitational binding, leading to a negative contribution to the overall binding energy. It is only due to the decreasing pressure in the outer regions that the overall binding energy is positive. One may therefore hope that the overall binding energy of a star composed of a pure quark core (also having a negative contribution to BE) and a hadronic shell (with its positive contribution to BE) turns out to be positive. Such neutron stars are conventionally called *hybrid stars* to emphasize the presence of quark matter (cf. Fig. 6.14).

Constant-pressure phase transition

The eQPM+l is intrinsically formulated for electrically neutral matter. Also, the neutron star EOSs are given for neutral matter. Thus, instead of global electric neutrality, a hybrid EOS of the two phases obeys – more restrictively – local electric neutrality. This, however, necessarily leads to a constant-pressure phase transition [Gle97] (i.e. a first-order phase transition) when requiring Gibbs conditions. In order to employ a rigorous variable-pressure construction, where only global electric neutrality is enforced and the pressure increases in a mixed phase region, it is necessary to entirely reformulate the quasiparticle model to include non-neutral matter. This would include answering questions on whether the electric charge of the quark-gluon-lepton medium should be included at $\mu = 0$ and carried along the characteristics to $T = 0$ or rather appear at $T = 0$ only, and whether lattice results are available for non-neutral matter.

6.9 Hybrid approaches

In a constant pressure phase transition no mixed phase appears: due to the monotonically decreasing pressure the coexistence region of both phases is reduced to a single radius of the quark star. Since the energy and the net quark density at equal pressure are, in general, not equal for two equations of state, both quantities are discontinuous at the phase transition and thus also in the star, as a sign of the implemented first order phase transition.

For the hadronic sector we make use of a common neutron star EOS, dubbed FPS by the authors [LRP93], which is derived from a Skyrme-like energy density functional fitted to the free energy of uniform-density nuclear and neutron matter as calculated by Friedman and Pandharipande [FP81]. For reference, we also use a neutron star EOS which displays a comparatively high energy density as a function of the pressure as it considers crystalline structures within the higher density phase [CC74] (dubbed EOS G as in the review [AB77]). As in Chapter 5 we use $n_q = 3n_{Ba}$ and thus $\mu = \mu_u = \mu_{Ba}/3$ to connect the quantities of hadronic and quark EOS.

The EOS $e(p)$ and the relations $n(p)$ as well as the thermodynamic potential[8], i.e. the pressure p/μ^4, are shown in Figs. 6.15 and 6.16. The hybrid EOS of the constant-pressure phase transition between the neutron star EOS FPS and the eQPM+l adjusted to [Bor10b] lattice results is shown as well. The pressure p_X at the phase transition is 2.236 GeV/fm^3. This somewhat larger value is the result of the rather small value of the pressure.

A series of similar constructions have been performed, e.g. exchanging the neutron star EOS FPS by EOS G or the quark part with the eQPM+l result when adjusting to the lattice results for the p4 action with $N_\tau = 8$ [Baz09] (where the value of p_X is only slightly smaller). Although not shown explicitly, these other hybrid EOS can be easily derived from the figures by determining p_X from the intersection of the two separate EOS in Fig. 6.15 and switching from $e(p)$ and $n(p)$ of the one to the other EOS at this p_X in Fig. 6.16.

Regardless of the EOS used for the outer neutron star shells, the existence of quark matter in the core drastically changes the relation of star mass M and radius R, as is visible in Fig. 6.17. As long as the central pressures of a hybrid star are below the transition pressure p_X no quark matter is present and the mass-radius relation follows the mass-radius relation of the neutron star. In stars with central pressures larger than p_X the core is made of deconfined matter, described by the eQPM+l EOS.

A close inspection of the TOV integration shows that, for hybrid stars in comparison to a FPS neutron star with the same central pressure p_c, the contained mass $m(r)$ rises faster with increasing radius until the phase transition is reached. The gained mass advantage is then carried to the surface, where the hybrid turns out slightly heavier and somewhat larger than the comparable neutron star. For hybrid stars with central pressures only a little larger than p_X this effect is barely noticeable due to the small quark cores but increases with enlarging quark content. As a result, the mass-radius curve is indented at the mass and radius of a neutron star with central pressure p_X, where the new (hybrid) branch with rapidly falling masses emerges.

---

[8]Constructed from given relations $e(p)$ and $n(p)$ at $T = 0$ by inverting $\mu = (e(p) + p)/n(p)$.

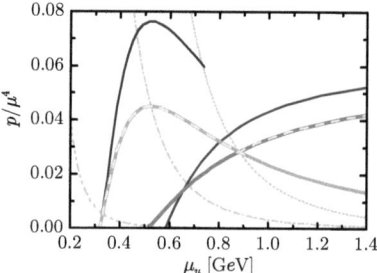

Figure 6.15: Scaled pressure p/μ^4 of the neutron star EOS FPS (yellow curve) and the eQPM+l adjusted to [Bor10b] lattice results (green curve) as well as the hybrid EOS of a constant-pressure phase transition (grey dashed curve). The grey dotted curve is the line of constant transition pressure $p_X = 2.236$ GeV/fm^3 while the grey dashed-dotted and dash-dot-dotted curves indicate the lines of constant pressure $p_X = 0.008$ GeV/fm^3 and $p_X = 0.440$ GeV/fm^3, respectively, for the generalized constant pressure transitions. For comparison, the scaled pressure of the neutron star EOS G (dark red curve) and the eQPM+l adjusted to [Baz09] lattice results with p4 action and $N_\tau = 8$ (blue curve) are shown. Note that due to the scaling curves with negative slopes still represent quantities increasing with the chemical potential as long as the slope is larger than the slopes of the lines of constant pressure.

Due to the generally small quark cores resulting from the constant-pressure phase transition, the large hadronic shell is able to contain the unbound core (i.e. the positive binding energy of the shell is larger than the absolute value of the negative binding energy of the core) and the hybrids would be bound (cf. Fig. 6.17). Since however, the bending, where the hybrids branch from the pure neutron star mass-radius curve, is on the left side of the cusp[9] of the neutron star mass-radius curve, these hybrid represent unstable configurations.

---

[9]This is a counterclockwise passed extremum of the mass-radius curve, where an unstable radial mode arises, cf. [Bar66, Har75, Kam81a].

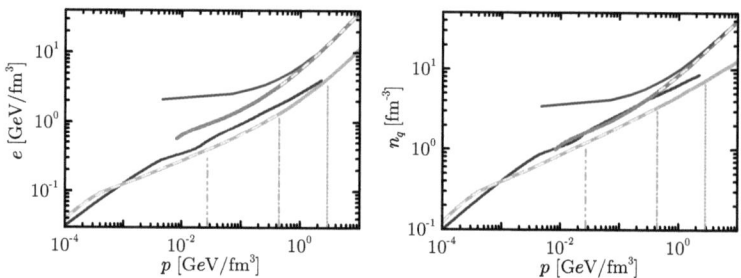

Figure 6.16: EOS $e(p)$ (left panel) and relation $n(p)$ (right panel) matching the scaled thermodynamic potentials p/μ^4 in Fig. 6.15 (same color coding and curve styles as therein). At the transition pressures p_X (indicated by the grey lines) jumps in both the energy density and the net particle density occur. For the neutron star EOS G these jumps are smaller.

Constant-pressure interpolation

It therefore suggests itself to, in a purely exploratory study, construct hybrid stars with smaller p_X so that the branching point of neutron and hybrid stars is located on the right side of the cusp. For this, one may relax the the requirement of one fixed chemical potential μ_X and, while still requiring the pressure to be constant in both phases, allow an interpolation between the two phases to start at one chemical potential μ_H and end at another chemical potential μ_Q [Kam94]. The pressure remains constant for all chemical potentials in the interval $[\mu_H, \mu_Q]$ (we adopt the denomination of hadronic/quark phase quantities using the indices H and Q, respectively).

While p_X was fixed in the former case due to the requirement of equal chemical potentials, this allows to select a range of p_X with, since μ_H and μ_Q are following from the choice, the only requirement being that $\mu_H(p_X) < \mu_Q(p_X)$. The pressure of the hybrid equation of state, as function of the chemical potential μ, is given by the hadron pressure $p_H(\mu)$ for $\mu \leq \mu_H$, where $p_H(\mu =: \mu_H) = p_X$, followed by the constant pressure p_X for all chemical potentials $\mu_H < \mu \leq \mu_Q$, where $p_Q(\mu =: \mu_Q) = p_X$, and for $\mu > \mu_Q$ by the quark pressure $p_Q(\mu)$. One has to note that this interpolation indeed only serves as an ad hoc tool to investigate the relation between various hybrid equations of state and the resulting stars on a phenomenological level, as, in the hybrid phase, the constant pressure p_X is strictly lower than the component pressures p_H and p_Q.

For two sensible values of p_X the interpolation is shown in Fig. 6.15 (note the scaling with μ^4). The first case is the minimum solution, i.e. $p_X = \min(p_Q)$, leading to a maximum quark core size, while, for the second case, p_X was chosen at the maximum of p/μ^4. We already considered the third case which is the constant-pressure phase transition at one fixed chemical potential $\mu_X = \mu_H = \mu_Q$ as the maximum solution of such a constant-pressure interpolation. Although not shown explicitly, the resulting hybrid EOS can be inferred from Fig. 6.16 using the vertical lines of constant pressure as guide.

As intended, the branching of neutron stars and hybrid stars constructed in this way occurs at larger radii (cf. Fig. 6.17). However, even though the objects are still mostly bound in the $p_X = 0.440$ GeV/fm^3 case, i.e. the hadronic shell is still large enough, the bending takes the form of a new cusp thus also leading to unstable configurations.

On the other hand, in the minimal case with largest possible quark cores a so-called third island of stability, i.e. a new region of stable configurations distinct from the neutron stars with smaller extents at equal masses, develops by means of a second cusp. As known from previous studies [Kam81a, Kam81b, Kam83] the stability of a neutron star with a discontinuous EOS and existence of a potential third island of stability is dependent on the jump of the (energy) density

$$\lambda_{\text{jump}} := \frac{e_Q}{e_H} \qquad (6.31)$$

and the value of the radius of the density jump. For the constant-pressure phase transition λ turns out to be 3.01 while it is about 2.36 for the latter interpolation, suggesting the existence

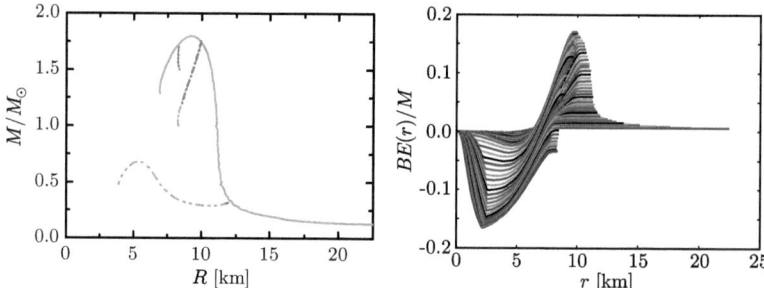

Figure 6.17: Left panel: Mass-radius relation of a neutron star (yellow curve) and several hybrid stars (green and grey curves). The curves match as long as the central pressure is too small to contain quark matter. The dotted curve represents the result for hybrid stars with constant-pressure phase transition (transition pressure $p_X = 2.236$ GeV/fm^3), while the dashed-dotted and dash-dot-dotted curves indicate the results for hybrids constructed from ad hoc constant-pressure interpolations with pressures $p_X = 0.008$ GeV/fm^3 and $p_X = 0.440$ GeV/fm^3 in the hybrid phase, respectively. At top neutron star masses the central pressure is 0.862 GeV/fm^3. The regions of bound configurations are colored while unbound regions are described as grey curves. Right panel: Contribution of inner shells to the scaled binding energy $BE(r)/M$ up to the radius r of a hybrid star with $p_X = 0.440$ GeV/fm^3. The end of each curve exhibits the overall binding energy $BE = BE(R)$. Even for pure neutron stars (low central pressures, large radii), the inner layers would be unbound if not surrounded by more dilute outer layers. For central pressures larger than p_X, the border between the quark core and the hadronic shell is clearly visible by the large negative contributions of the former to the binding energy. As long as the quark core is light enough, the outer layers also provide binding to the latter. Clearly, there are configurations with very massive (and even again smaller) cores which cannot be contained.

of such a third island. However, for these stable configurations, the hadronic shell is only about 500 m thick[10] and not able to bind the quark core. In fact, on this branch with smallest possible hadronic shells, binding already stops at very large radii.

All prior investigations were also carried out using the EOS G for the hadronic part and/or the eQPM+l EOS when adjusted to the lattice results using the p4 action and $N_\tau = 8$ [Baz09]. The results were largely equivalent, especially if comparing the constant-pressure phase transition using the two eQPM+l results, where λ turns out to be nearly identical. Consequently, we can conclude our exploration with the assessment that, if one trusts the EOS obtained from the most recent lattice results using the eQPM+l and a constant-pressure phase transition or interpolation, the existence of pure quark stars as well as neutrons stars with a quark core is not suggested, as the found configurations were either unstable or unbound.

Variable-pressure interpolation

The limitation to constant-pressure phase transitions, as result of the local charge neutrality, restricts the transition region within the hybrid star to one single radius. In order to, at least phenomenologically, simulate the effects of a variable-pressure phase transition and check if a more advanced structure of the phase transition and a broad transition region would allow for stable and simultaneously bound hybrid configurations, we investigate a variable-pressure interpolation[11].

One may model the continuous increase of energy and net quark density in the mixed phase between two values of $p_x^H < p_x^Q$[12] limiting the phase transition region by a parametrization $e = e_M(p)$ and $n = n_M(p)$ with $e_M(p_x^{H/Q}) = e_{H/Q}(p_x^{H/Q})$ and $n_M(p_x^{H/Q}) = n_{H/Q}(p_x^{H/Q})$, i.e. the nonzero first derivatives of the pressure are continuous. Since, due to the parametrization of the transition region, the second derivatives are in general not continuous, this imitates a second order phase transition. In order to verify the thermodynamic consistency of the approach one has to ensure that the resulting chemical potential $\mu = (e+p)/n$ is a strictly monotonic increasing as function of the pressure $p\epsilon(p_x^H, p_x^Q)$.

The most accessible nontrivial ansatz ensuring the latter is to assume a linear dependency of energy and net quark density on the pressure, i.e. $e_M = p(e_Q(p_X^Q) - e_H(p_X^H))/(p_X^Q - p_X^H) + e_H(p_X^H)$ and equivalent for the net particle density. For this investigation we use the quark EOS found using the eQPM+l adjusted to [Baz09] lattice results using the p4 action with $N_\tau = 8$.

---

[10]It is interesting to note that for all stable configurations in the third island this absolute size of the hadronic shell is almost constant and seemingly independent of the size of the quark core.
[11]Similar investigations using a simple bag model for the quark phase have been presented in [Cha08, Bha10].
[12]We may exclude the limiting case $p_x^H = p_x^Q$ as it is equal to a constant-pressure interpolation.

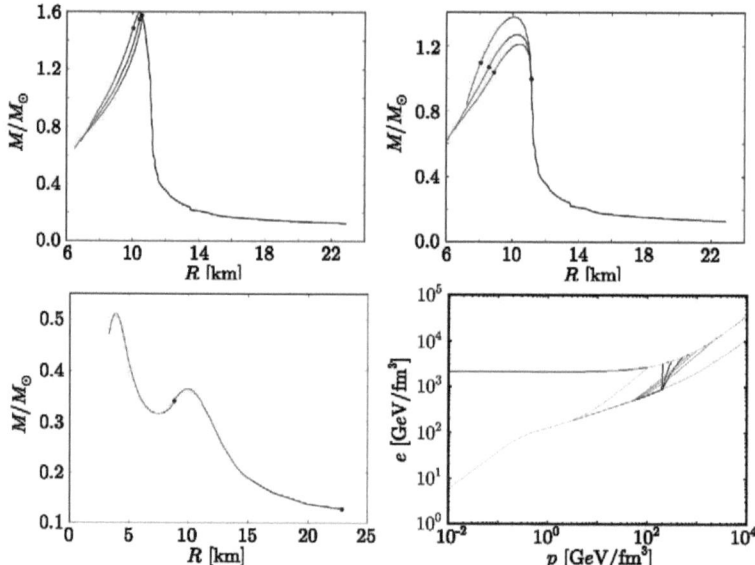

Figure 6.18: Mass-radius relations $M(R)$ of several hybrid stars for which a variable pressure in an extended transition region is simulated. Blue curves represent bound configurations while stars on the red curves are unbound. The blue dots indicate the points where the central pressure equals p_X^H (right dot) and p_X^Q (left dot). The core of configurations left/between/right of the two dots contains pure quark/hybrid/hadronic matter, respectively. Top left panel: Changes from a constant-pressure to a broad variable-pressure interpolation for values $p_X^H = 0.2$ GeV/fm^3 and $p_X^Q = [0.2, 0.3, 0.5]$ GeV/fm^3. Top right panel: Stable and bound hybrid stars resulting from the construction from $p_X^H = 0.05$ GeV/fm^3 to $p_X^Q = [0.7, 1.0, 2.0]$ GeV/fm^3. Bottom left panel: For configurations with low p_X^H (here 0.002 GeV/fm^3 matched to $p_X^Q = 0.1$ GeV/fm^3) the hybrid branch is unbound even to the right side the cusp. A (also unbound) third island visible. Bottom right panel: Equations of state $e(p)$ considered in the other three panels (black curves: top left, dark grey curves: top right, light grey curves: bottom left) as variable-pressure interpolation between the eQPM+l EOS adjusted to lattice results using the p4 action with $N_\tau = 8$ from [Baz09] (blue curve) and the neutron star EOS FPS (yellow curve).

Indeed, stable and simultaneously bound hybrid stars can be obtained using this construction. As to be expected from the previous subsection, the cusp of the mass-radius curves always occurs at central pressures smaller than p_X^Q so that the quark matter in the stable configurations is always contained in the hybrid phase rather than a pure quark phase. If $p_X^Q \to p_X^H$ the cusp turns into an indentation of the neutron star mass-radius relation at $p_X^Q = p_X^H$. Although variable-pressure interpolated hybrids would be possible for p_X^H up to 0.862 GeV/fm^3, i.e. the pressure at the cusp of the neutron star mass-radius curve, bound configurations are found only for $p_X^H < 0.7$ GeV/fm^3. Also, for small values of p_X^H, i.e. early inset of the interpolation to the quark phase, the stable hybrids quickly become unbound.

Changing our simple linear interpolation to a more sophisticated ansatz would introduce some minor numerical differences, however, no significant change of our purely qualitative results are to be expected. We may therefore conclude that, while no pure quark cores are suggested by the investigation, there may exist – depending of the character and broadness of the interpolation region (and thus of the phase transition region it models) – neutron stars containing deconfined quarks in a hybrid phase in the innermost layers.

6.10 Wrap-up

In Fig. 5.12 it is visible that the use of the HTL QPM instead of the eQPM leads to slightly increased state quantities at $T = 0$. This variation is especially small for the pressure (note the scaling factor). One may therefore assume that the extension of the HTL QPM, in the same way as done here for the eQPM, also leads to only small changes, too. At most the fact that the energy density increase is larger than the rise of the pressure would indicate a somewhat higher vacuum energy density e_0 but only an insignificant change in the inverse square of the velocity of sound v_s^{-2}. From this, qualitative changes of the results presented here are not to be expected. As our investigation bears a strictly explorative character, the study of such a HTL+l QPM, which would analytically and numerically be much more involved than the eQPM+l, for purely quantitative improvements was not put forward.

Of course, in general, the applicability of such models at low temperatures and high densities is not guaranteed. As shown, the mass-radius relations of ordinary neutron stars and neutron stars with quark cores are quite similar in the region where the latter are bound. This renders an experimental verification via these observables difficult. The modified cooling behavior of quark matter is considered as a possible tool to find appropriate observational hints [Bla00]. Our result that the existence of pure quark stars is extremely unlikely is in line with the predictions by the group of J. Lattimer [Pos10] using a different approach.

7 Summary

In this thesis, the full Hard Thermal Loop (HTL) quasiparticle model (QPM) was introduced as a thermodynamic model of deconfined matter, following from QCD via the Cornwall-Jackiw-Tomboulis formalism and a series of approximations and assumptions. Due to thermodynamic self-consistency the model allows for an extrapolation of lattice results, available at vanishing chemical potential and well described due to a special parametrization of the running coupling, to large baryon densities.

In contrast to the established effective quasiparticle model (eQPM), quasiquarks and -gluons obey implicit dispersion relations with energy- and momentum-dependent self-energies and quasiparticles as quanta of collective excitations (plasmons and plasminos) appear as well as Landau damping below the light cone. The pure quasiparticle contributions of the collective modes to pressure, entropy and net particle density were shown to be negative. As a result, the extrapolation of the lattice results yields a region of negative pressure indicating a necessary transition to a different state of matter. A transition connecting the HTL quasiparticle model to the hadron resonance gas was considered providing an equation of state (EOS) for experiments at SPS. The construction of an EOS for compressed-baryon-matter experiments at FAIR turns out to be not sensible due to largely different isentropic curves in the confined and deconfined phases for large entropies per baryon. Changes due to the inclusion of charm quarks are not expected.

Comparison of the HTL QPM to the eQPM revealed a remarkable similarity of the extrapolated thermodynamic state variables. As a result, the analytically easier accessible eQPM was used in an exploratory study aimed to include the weakly interacting sector into the model in order to describe β-stable quark matter in compact stellar objects. We derived general scaling laws and binding arguments allowing to connect properties of the EOS to statements concerning the possible existence of pure quark stars. These were confirmed in the investigation of pure quark star properties for the EOS derived from recent lattice results using the QPM extrapolation procedure. From the results, the existence of absolutely stable and bound pure quark stars is not suggested. One may argue towards a possible existence as meta-stable compact stellar objects.

Finally, the possible existence of deconfined matter in neutron star cores ("hybrid stars") was investigated on a phenomenological level. If the transition from the neutron star EOS to the quark EOS is a constant-pressure phase transition, i.e. neutron and quark phase are strictly separated, our studies suggest no stable hybrids. For a variable-pressure phase transition, where a mixed phase appears, stable and at the same time bound configurations were found and might be realized. All found third islands of stability turn out to be unbound.

Appendix A Evaluation of Matsubara sums

Considering an arbitrary function $f(p_0 = i\omega_n)$ which is to be summed over all bosonic Matsubara frequencies $i\omega_n = 2ni\pi T$ one defines a quantity

$$M := T \sum_{n=-\infty}^{+\infty} f(p_0 = i\omega_n). \tag{A.1}$$

The factor T in front of the sum will turn out to be very helpful in performing the sum.

We start by examining the bosonic distribution function for imaginary frequencies $n_B(p_0 = i\omega) = (e^{i\beta\omega} - 1)^{-1}$. It has poles at the bosonic Matsubara frequencies with residuum[1] T

$$\operatorname{Res}_{i\omega_n} n_B = \operatorname{Res}_{i\omega_n} \frac{1}{e^{\beta p_0} - 1} = \frac{T}{e^{\beta i\omega_n}} = T, \tag{A.3}$$

so that M can be written as

$$M = \sum_n f(i\omega_n) \operatorname{Res}\left[n_B(p_0 = i\omega_n)\right]. \tag{A.4}$$

We now require $f(i\omega)$ to be analytic at the poles of the bosonic distribution function n_B and n_B to be analytic at the poles of $f(i\omega)$. As a consequence, the set of poles of the product $f(i\omega) n_B(i\omega)$ is equal to the union of the two sets of poles of $f(i\omega)$ and $n_B(i\omega)$:

$$\{i\omega_m\} = \{i\omega_n\} + \{i\omega_l\}. \tag{A.5}$$

Here m counts the poles of the product and l and n those of $f(i\omega)$ and $n_B(i\omega)$, respectively. This implies the relation

---

[1] For functions $f(z)$ with a pole of first order at $z = a$ we find

$$\operatorname{Res}_a \frac{1}{f(z)} = \lim_{z \to a} (z-a) \frac{1}{f(z)} \stackrel{f(a)=0}{=} \lim_{z \to a} \frac{z-a}{f(z) - f(a)} =: \frac{1}{f'(a)}. \tag{A.2}$$

$$\sum_m \text{Res}\left[n_B(i\omega_m)f(i\omega_m)\right] = \sum_l \text{Res}\left[f(i\omega_l)\right] n_B(i\omega_l) + \underbrace{\sum_n f(i\omega_n) \text{Res}\left[n_B(i\omega_n)\right]}_{=M}. \quad (A.6)$$

Each of these terms can be translated back into a contour integral encircling the respective poles. In particular, the first integral then represents the contour integral containing all poles. Its contour can be moved to infinity, where the integral vanishes as long as $f(i\omega)$ is a monotonically decreasing function going to zero for $\omega \to \pm\infty$.[2] This leaves us with

$$M = -\sum_l \text{Res}\left[f(i\omega_l)\right] n_B(\omega = i\omega_l). \quad (A.7)$$

Let us reintroduce a complex energy integration, so that

$$M = -\frac{1}{2\pi} \int_{-i\infty}^{+i\infty} d(i\omega') \sum_l 2\pi\delta(i\omega' - i\omega_l) \text{Res}[f(i\omega')] n_B(i\omega'). \quad (A.8)$$

After substituting $i\omega' \to \omega$ we find the well-known structure of a spectral density $\varrho_f(\omega) := \sum_l 2\pi\delta(\omega - \omega_l) \text{Res}_{\omega_l} f(\omega)$

$$M = -\frac{1}{2\pi} \int_{-\infty}^{+\infty} d\omega \underbrace{\sum_l 2\pi\delta(\omega - \omega_l) \text{Res}\left[f(\omega)\right]}_{=:\varrho_f(\omega)=2\text{Im}\,f(\omega+i\varepsilon)} n_B(\omega) \quad (A.9)$$

which can be replaced by the imaginary part of the retarded expression[3] for f, finally yielding

---

[2] While the expression is Boltzmann suppressed by $n_B(i\omega)$ for $\omega \to \pm i\infty$, since $n_B(i\omega) \overset{i\omega \to \pm\infty}{\longrightarrow} e^{\mp\beta\omega}$, it is purely oscillatory for $\omega \to \pm\infty$ and does not interfere with the asymptotics of $f(i\omega)$.

[3] As an example, this is shown for a bosonic propagator $f(\omega) = D(\omega) = -(\omega^2 - \omega_k^2)^{-1}$. The expression

$$\text{Res}_{\pm\omega_k} D(\omega) = \lim_{\omega \to \pm\omega_k} \frac{-(\omega \mp \omega_k)}{(\omega + \omega_k)(\omega - \omega_k)} = \lim_{\omega \to \pm\omega_k} \frac{-1}{\omega \pm \omega_k} = \mp\frac{1}{2\omega_k}$$

leads to $\varrho_D(\omega) = 2\pi\left[\delta(\omega + \omega_k) - \delta(\omega - \omega_k)\right]/(2\omega_k)$. Using $(A + i\varepsilon)^{-1} = PA^{-1} - i\pi\delta(A)$, where P denotes the principal value, we obtain the same result for

$$2\text{Im}\,D(\omega + i\epsilon) = -2\text{Im}\,\frac{1}{\omega^2 + 2i\epsilon\omega - \omega_k^2}$$
$$= 2\pi\delta\left(\omega^2 - \omega_k^2\right) = \frac{2\pi}{2\omega_k}\left[\delta(\omega + \omega_k) - \delta(\omega - \omega_k)\right]$$

In fact, this simple example can be expanded to a rigorous proof by treating an arbitrary spectral function as infinite sum of delta distributions (convolution integral) and calculating its propagator using the Lehmann representation.

$$M = -\int_{-\infty}^{+\infty} \frac{d\omega}{\pi} n_B(\omega) \operatorname{Im}(f(\omega + i\varepsilon)). \qquad (A.10)$$

For a sum over all fermionic Matsubara frequencies, one replaces n_B by the Fermi-Dirac distribution function n_F with poles at $i\omega_n = (2n+1)i\pi T + \mu$ and $\operatorname{Res}_{i\omega_n} n_F = T$ in Eq. (A.4). Since the remaining steps are independent of the explicit form of the distribution function, the result is found in an analogous way.

Appendix B Mathematical relations

B.1 Imaginary part of the logarithm

The imaginary part of the logarithm of a complex quantity z (e.g. an inverse propagator) equals the argument of z as

$$\begin{aligned}
\operatorname{Im}\ln(z) &= \operatorname{Im}\ln\left(|z|e^{i\operatorname{Arg}(z)}\right) \\
&= \operatorname{Im}\bigl(\underbrace{\ln|z|}_{\in\mathbb{R}} + i\operatorname{Arg}(z)\bigr) \\
&= \operatorname{Arg}(z).
\end{aligned} \qquad (\text{B.1})$$

Therefore, z is allowed to have a dimension, e.g. in the case of $z = D_T^{-1}$ a (squared) energy dimension, even though the logarithm itself is defined for dimensionless numbers only. For explicit calculations, the dimension has to be removed:

$$\operatorname{Im}\ln D_T^{-1} = \operatorname{Im}\Bigl(\ln\frac{D_T^{-1}}{T^2} + \underbrace{2\ln T}_{\in\mathbb{R}}\Bigr) = \operatorname{Im}\ln\frac{D_T^{-1}}{T^2}. \qquad (\text{B.2})$$

This is not an ambiguity since it has no influence on the argument of D_T^{-1}. For instance, the argument can be calculated using the arc tangent. Compensating for quadrant relations it is given by

$$\operatorname{Im}\ln(z) = \operatorname{Arg}(z) = \arctan\frac{\operatorname{Im}z}{\operatorname{Re}z} + \pi\varepsilon(\operatorname{Im}z)\Theta(-\operatorname{Re}z) \qquad (\text{B.3})$$

and, if the argument is $-z$,

$$\operatorname{Im}\ln(-z) = \operatorname{Arg}(-z) = \arctan\frac{\operatorname{Im}z}{\operatorname{Re}z} - \pi\varepsilon(\operatorname{Im}z)\Theta(\operatorname{Re}z). \qquad (\text{B.4})$$

Note that the step function Θ is defined for dimensionless quantities only too, so implicitly it is always to be divided by a reference quantity (e.g. again a power of the temperature if z is an inverse propagator).

B.2 Derivative of Arg and arctan

Starting with the derivative of the Heaviside function of a function x/a (see footnote 3 in Section 3.2)

$$\frac{\partial}{\partial x}\Theta\left(\frac{x}{a}\right) = \frac{1}{a}\delta\left(\frac{x}{a}\right) = \frac{|a|}{a}\delta(x) = \varepsilon(a)\delta(x) \tag{B.5}$$

we find the derivative of the sign function $\varepsilon(x) := \Theta(x) - \Theta(-x)$

$$\frac{\partial}{\partial x}\varepsilon(\frac{x}{a}) = 2\varepsilon(a)\delta(x). \tag{B.6}$$

Using a symmetry relation and the derivative of the arc tangent (cf. [TBM01])

$$\arctan\left(\frac{a}{x}\right) = -\arctan\left(\frac{x}{a}\right) + \pi\Theta\left(\frac{x}{a}\right) - \frac{\pi}{2}, \tag{B.7}$$

$$\frac{\partial}{\partial x}\arctan\left(\frac{x}{a}\right) = \frac{a}{a^2 + x^2} \tag{B.8}$$

we find

$$\frac{\partial}{\partial x}\arctan\left(\frac{a}{x}\right) = -\frac{a}{a^2 + x^2} + \pi\varepsilon(a)\delta(x). \tag{B.9}$$

Applying this relation to the argument $\text{Arg}\, z = \arctan(\text{Im}\, z/\text{Re}\, z) + \pi\varepsilon(\text{Im}\, z)\Theta(-\text{Re}\, z)$ of a complex quantity z (e.g. an inverse propagator) we find

$$\begin{aligned}
\frac{\partial}{\partial \text{Re}\, z}\text{Arg}\, z &= \frac{-\text{Im}\, z}{\text{Im}^2 z + \text{Re}^2 z}, \\
\frac{\partial}{\partial \text{Im}\, z}\text{Arg}\, z &= \frac{\text{Re}\, z}{\text{Im}^2 z + \text{Re}^2 z} + 2\pi\delta(\text{Im}\, z)\Theta(-\text{Re}\, z).
\end{aligned} \tag{B.10}$$

The two emerging Dirac delta distributions of the derivative with respect to the real part of z exactly cancel.

Appendix C List of derivatives

C.1 Derivatives of the HTL thermal masses

The derivatives of the Debye mass m_D^2 are

$$\left.\frac{\partial m_D^2}{\partial T}\right|_{\mu,G^2} = \frac{2}{3}C_b TG^2 \quad \text{and} \quad \left.\frac{\partial m_D^2}{\partial T}\right|_{\mu} = \left.\frac{\partial m_D^2}{\partial T}\right|_{\mu,G^2} + \left.\frac{\partial m_D^2}{\partial G^2}\right|_{T,\mu} \frac{\partial G^2}{\partial \mu},$$

$$\left.\frac{\partial m_D^2}{\partial \mu}\right|_{T,G^2} = \frac{N_c N_l}{3\pi^2}\mu G^2 \quad \text{and} \quad \left.\frac{\partial m_D^2}{\partial \mu}\right|_{T} = \left.\frac{\partial m_D^2}{\partial \mu}\right|_{T,G^2} + \left.\frac{\partial m_D^2}{\partial G^2}\right|_{T,\mu} \frac{\partial G^2}{\partial \mu},$$

$$\left.\frac{\partial m_D^2}{\partial G^2}\right|_{T,\mu} = 2\tilde{C}_b \qquad (\text{C.1})$$

while the derivatives of the fermionic mass parameter/plasma frequency $\hat{M}^2$ of the light quarks are

$$\left.\frac{\partial \hat{M}^2}{\partial T}\right|_{\mu,G^2} = \frac{C_f}{4}TG^2 \quad \text{and} \quad \left.\frac{\partial \hat{M}^2}{\partial T}\right|_{\mu} = \left.\frac{\partial \hat{M}^2}{\partial T}\right|_{\mu,G^2} + \left.\frac{\partial \hat{M}^2}{\partial G^2}\right|_{T,\mu} \frac{\partial G^2}{\partial \mu},$$

$$\left.\frac{\partial \hat{M}^2}{\partial \mu}\right|_{T,G^2} = \frac{C_f}{4\pi^2}\mu G^2 \quad \text{and} \quad \left.\frac{\partial \hat{M}^2}{\partial \mu}\right|_{T} = \left.\frac{\partial \hat{M}^2}{\partial \mu}\right|_{T,G^2} + \left.\frac{\partial \hat{M}^2}{\partial G^2}\right|_{T,\mu} \frac{\partial G^2}{\partial \mu},$$

$$\left.\frac{\partial \hat{M}^2}{\partial G^2}\right|_{T,\mu} = \tilde{C}_f. \qquad (\text{C.2})$$

The combined derivatives to the right are needed e.g. for the integration of the mean field pressure along the T or μ axis.

The second derivatives are given by

$$\left.\frac{\partial^2 \hat{M}^2}{\partial T^2}\right|_{\mu} = \frac{C_f}{8}\left(2G^2 + 4T\frac{\partial G^2}{\partial T} + \left(T^2 + \frac{\mu^2}{\pi^2}\right)\frac{\partial^2 G^2}{\partial T^2}\right),$$

$$\left.\frac{\partial^2 \hat{M}^2}{\partial \mu^2}\right|_{T} = \frac{C_f}{8}\left(\frac{2}{\pi}G^2 + \frac{4\mu}{\pi^2}\frac{\partial G^2}{\partial \mu} + \left(T^2 + \frac{\mu^2}{\pi^2}\right)\frac{\partial^2 G^2}{\partial \mu^2}\right).$$

Due to $\hat{M}_s^2 = \hat{M}^2|_{\mu=0}$, the derivatives for the strange quark follow by setting $\mu = 0$ in the above expressions.

C.2 Derivatives of the eQPM asymptotic masses

Effective quasiparticle model without leptons

The derivatives of the asymptotic masses $\tilde{m}_{i,\infty}^2$ with temperature restmasses $m_{i,0}^2$ as defined in Eqs. (3.6) and (2.19) with $\mu = \mu_u = \mu_d$ and $\mu_s = 0$ are given here. Taking the derivative of the $\tilde{m}_{i,\infty}^2$ with respect to temperature T, chemical potential μ or effective coupling G^2, while keeping the respective other two quantities constant (which we imply when using the squared brackets $[\,]$), we have

$$\frac{\partial \tilde{m}_{i,\infty}^2}{\partial [T|\mu|G^2]} = 2\left(m_{i,0} + \sqrt{2}m_{i,\infty}\right)\frac{\partial m_{i,0}}{\partial [\,]} + \left(\frac{m_{i,0}}{\sqrt{2}m_{i,\infty}} + 1\right)\frac{\partial m_{i,\infty}^2}{\partial [\,]}, \quad (C.3)$$

where a possible dependence of the restmasses on T, μ or G^2 can be accommodated. This has been done to compare to previous lattice calculations with lattice restmasses $m_{q,0} = aT$, e.g. from [Pei00].[1]

The derivatives of the asymptotic masses $m_{i,\infty}^2$ follow directly from the derivatives of the HTL thermal masses. From Eqs. (3.3) and (3.5) we have

$$\frac{\partial m_{g,\infty}^2}{\partial [\,]} = \frac{1}{2}\frac{\partial m_D^2}{\partial [\,]} \quad \text{and} \quad \frac{\partial m_{q,\infty}^2}{\partial [\,]} = 2\frac{\partial \hat{M}^2}{\partial [\,]}. \quad (C.4)$$

The second derivatives are given by

$$\frac{\partial^2 \tilde{m}_{i,\infty}^2}{\partial [\,]^2} = 2\left(\frac{\partial m_{i,0}}{\partial [\,]}\right)^2 + \frac{\sqrt{2}}{m_{i,\infty}}\frac{\partial m_{i,\infty}^2}{\partial [\,]}\frac{\partial m_{i,0}}{\partial [\,]} + 2m_{i,0}\frac{\partial^2 m_{i,0}}{\partial [\,]^2} + \sqrt{2}m_{i,\infty}\frac{\partial^2 m_{i,0}}{\partial [\,]^2}$$
$$+ \sqrt{2}m_{i,0}\left(\frac{1}{2m_{i,\infty}}\frac{\partial^2 m_{i,\infty}^2}{\partial [\,]^2} - \frac{1}{4m_{i,\infty}^3}\left(\frac{\partial m_{i,\infty}^2}{\partial [\,]}\right)^2\right) + \frac{\partial^2 m_{i,\infty}^2}{\partial [\,]^2} \quad (C.5)$$

with $\partial^2 m_{i,\infty}^2/\partial [\,]^2$ following also from the HTL expressions.

The explicit expression for the gluons is

$$\frac{\partial^2 \tilde{m}_{g,\infty}^2}{\partial T^2} = \frac{C_b}{3}G^2 + 2\frac{C_b}{3}T\frac{\partial G^2}{\partial T} + \tilde{C}_b\frac{\partial^2 G^2}{\partial T^2}.$$

The strange quark expressions follow again by setting $\mu = 0$.

Effective quasiparticle model with leptons

The derivatives of the asymptotic masses $\tilde{m}_{i,\infty}^2$ with temperature-independent restmasses $m_{i,0}^2$ as defined in Section 6.3 with $\mu = \mu_u$ and $\mu_d = \mu_s = \mu + \mu_e$ are given here. For the gluons we

---
[1]This thesis does, however, not report on any of the results, as the lattice data are outdated.

C.2 Derivatives of the eQPM asymptotic masses

obtain

$$\left.\frac{\partial m_{g,\infty}^2}{\partial T}\right|_{G^2,\mu} = \left(\frac{C_b}{3}T + \frac{N_c}{12\pi^2}(4\mu + 4\mu_e)\left.\frac{\partial \mu_e}{\partial T}\right|_{G^2,\mu}\right)G^2,$$

$$\left.\frac{\partial m_{g,\infty}^2}{\partial \mu}\right|_{G^2,T} = \frac{N_c}{12\pi^2}\left((6\mu + 4\mu_e) + (4\mu + 4\mu_e)\left.\frac{\partial \mu_e}{\partial \mu}\right|_{G^2,T}\right)G^2,$$

$$\left.\frac{\partial m_{g,\infty}^2}{\partial G^2}\right|_{T,\mu} = \frac{C_b}{6}T + \frac{N_c}{12\pi^2}(4\mu + 4\mu_e)\left.\frac{\partial \mu_e}{\partial G^2}\right|_{T,\mu}. \qquad (C.6)$$

The derivative with respect to the chemical potential in Eq. (C.3) is modified due to the additional dependence of μ_d and μ_s on the electron chemical potential μ_e. For $q\epsilon\{d, s\}$ we have

$$\frac{\partial \tilde{m}_{q,\infty}^2}{\partial \mu} = \left(2\left(m_{i,0} + \sqrt{2}m_{i,\infty}\right)\frac{\partial m_{i,0}}{\partial \{\}} + \left(\frac{m_{i,0}}{\sqrt{2}m_{i,\infty}} + 1\right)\frac{\partial m_{i,\infty}^2}{\partial \{\}}\right)\left(1 + \frac{\partial \mu_e}{\partial \mu}\right) \qquad (C.7)$$

while $\partial \tilde{m}_{q,\infty}^2/\partial T$ and $\partial \tilde{m}_{q,\infty}^2/\partial G^2$ are formally unmodified.

Appendix D Coefficients of the flow equations

From Eq. (2.55) we have the general definition

$$a_T = -\sum \frac{\partial n_i}{\partial \Pi_i} \frac{\partial \Pi_i}{\partial G^2},$$

$$a_\mu = \sum \frac{\partial s_i}{\partial \Pi_i} \frac{\partial \Pi_i}{\partial G^2},$$

$$b = \sum \frac{\partial n_i}{\partial \Pi_i} \frac{\partial \Pi_i}{\partial T}\bigg|_{G^2} - \frac{\partial s_i}{\partial \Pi_i} \frac{\partial \Pi_i}{\partial \mu}\bigg|_{G^2}, \quad \text{(D.1)}$$

where the self-energies Π_i have to be replaced by the asymptotic masses $\tilde{m}_{i,\infty}^2$ for the eQPM quasiparticle models. The necessary derivatives are given in the following.

D.1 Effective quasiparticle models

For the eQPM the sums run over $i = g, q$ and a possible strange quark s. The derivatives of the $\tilde{m}_{i,\infty}^2$ with respect to the effective coupling and the temperature and chemical potential when keeping G^2 constant are given in Appendix C. The derivatives of the contributions to the net quark density and the partial entropy densities with respect to the $\tilde{m}_{i,\infty}^2$ are

$$\frac{\partial n_i^{eQP}}{\partial \tilde{m}_{i,\infty}^2} = \frac{d_i}{4\pi^2 T} \int_0^\infty dk \, \frac{k^2}{\omega_{\text{TL}}(k)} \left[e^+ f_-^2 - e^- f_+^2 \right],$$

$$\frac{\partial s_i^{eQP}}{\partial \tilde{m}_{i,\infty}^2} = \frac{d_i}{4\pi^2 T^2} \int_0^\infty dk \, k^2 \left\{ -\left[e^- f_+^2 + e^+ f_-^2 \right] + \frac{\mu}{\omega_{\text{TL}}(k)} \left[e^- f_+^2 + e^+ f_-^2 \right] \right\}. \quad \text{(D.2)}$$

The quark degeneracy factors are $d_q = 2N_cN_l$ and $d_s = 2N_c$.

Formally, the derivatives of n_i and s_i are preserved when going from the eQPM to the eQPM+l model. The sums then run over $i = g, u, d, s$ with degeneracy factors $d_{\{u,d,s\}} = 2N_c$. The mass derivatives for the eQPM+l can also be found in Appendix C.

The derivatives of the coefficients at vanishing chemical potential needed in the Taylor expansion at vanishing chemical potential (cf. Section 4.5) are

$$\left.\frac{\partial b}{\partial \mu}\right|_{\mu=0} = \left.\frac{\partial \tilde{m}_{q,\infty}^2}{\partial T}\right|_{G^2} \frac{\partial^2 n_q}{\partial \mu \, \partial m_q^2} - \left(\frac{\partial}{\partial \mu} \left.\frac{\partial \tilde{m}_{q,\infty}^2}{\partial \mu}\right|_{G^2}\right) \frac{\partial s_q}{\partial m_q^2} - \left(\frac{\partial}{\partial \mu} \left.\frac{\partial \tilde{m}_{g,\infty}^2}{\partial \mu}\right|_{G^2}\right) \frac{\partial s_g}{\partial m_g^2},$$

$$\left.\frac{\partial a_T}{\partial \mu}\right|_{\mu=0} = -\left.\frac{\partial \tilde{m}_{q,\infty}^2}{\partial G^2}\right|_{T,\mu} \frac{\partial^2 n_q}{\partial \mu \, \partial m_q^2} \tag{D.3}$$

with

$$\left(\frac{\partial}{\partial \mu} \left.\frac{\partial \tilde{m}_{q,\infty}^2}{\partial \mu}\right|_{G^2}\right)_{\mu=0} = \left(\frac{m_{q,0}}{\sqrt{2} m_{i,\infty}} + 1\right) \frac{C_f}{2\pi^2} G^2\big|_{\mu=0}, \tag{D.4}$$

$$\left(\frac{\partial}{\partial \mu} \left.\frac{\partial \tilde{m}_{g,\infty}^2}{\partial \mu}\right|_{G^2}\right)_{\mu=0} = \frac{3 N_l}{6\pi^2} G^2\big|_{\mu=0} \tag{D.5}$$

and

$$\left.\frac{\partial^2 n_q^{eQP}}{\partial \mu \, \partial m_q^2}\right|_{\mu=0} = \frac{d_q}{2\pi^2 T} \int_0^\infty dk \, \frac{k^2}{\omega_{\mathrm{TL}}(k)} \left[ef^2 - 2e^2 f^3\right]. \tag{D.6}$$

D.2 HTL quasiparticle model

Note that in the following a somewhat different abbreviation scheme is used than for the effective quasiparticle models. This is due to a different structure of the derivatives of entropy density and particle density with respect to μ and T, respectively, since the HTL QPM uses the full dispersion relations as opposed to the asymptotic dispersion relations of eQPM.

Gluons

As first part of the Maxwell relation, the derivative of the HTL gluon entropy density $s_g = s_{g,\mathrm{T}} + s_{g,\mathrm{L}}$ with partial gluon entropy densities (2.36) with respect to μ at constant T has to be calculated. For gluons there is no explicit dependence of the entropy density on the chemical potential, so that only the self-energies and propagators depend on μ due to the contained Debye mass. Consequently, there are four contributions to the derivative

$$\frac{\partial s_g}{\partial \mu} = \left(\frac{\partial s_g}{\partial \mu}\right)_{\mathrm{Re}D_{\mathrm{T}}^{-1}} + \left(\frac{\partial s_g}{\partial \mu}\right)_{\mathrm{Im}\Pi_{\mathrm{T}}} + \left(\frac{\partial s_g}{\partial \mu}\right)_{\mathrm{Re}D_{\mathrm{L}}^{-1}} + \left(\frac{\partial s_g}{\partial \mu}\right)_{\mathrm{Im}\Pi_{\mathrm{L}}}, \tag{D.7}$$

where the index on the bracket indicates the considered dependence.

D.2 HTL quasiparticle model

For the first term we find

$$\left(\frac{\partial s_g}{\partial \mu}\right)_{\text{Re}D_T^{-1}} = -\frac{2d_g}{\pi m_D^2}\frac{\partial m_D^2}{\partial \mu}\int d^3k \int_0^\infty d\omega \frac{\partial n_B}{\partial T} \qquad (D.8)$$

$$\times \left\{-\frac{\text{Re}\Pi_T \text{ Im}\Pi_T}{\text{Re}^2 D_T^{-1} + \text{Im}^2 \Pi_T} + \text{Re}\Pi_T \text{Im}\Pi_T \frac{\text{Re}^2 D_T^{-1} - \text{Im}^2 \Pi_T}{(\text{Re}^2 D_T^{-1} + \text{Im}^2 \Pi_T)^2}\right\},$$

where the derivative of the quasiparticle pole term $\pi\Theta()$ is canceled by the term $\pi\delta()$ arising from the derivative of the arc tangent. We split the energy integration at the light cone. For $\omega > k$ the imaginary part of the transverse gluon self-energy is equal to $\eta\varepsilon(\text{Im}\Pi_T)$, where $\eta \to 0$ due to retardation (cf. Eqs. 2.27). Leaving aside the prefactor $-\text{Re}\Pi_T$ and after multiplication by $\varepsilon(\text{Im}\Pi_T)/\pi$ to assure a positive width $\Gamma := 2|\text{Im}\Pi_T|$ and normalization, the first term of the curly bracket corresponds to a Breit-Wigner distribution of a quantity $x = \text{Re}D_T^{-1}$:

$$\frac{1}{\pi}\frac{|\text{Im}\Pi_T|}{\text{Re}^2 D_T^{-1} + \text{Im}^2 \Pi_T} \longleftrightarrow \frac{1}{2\pi}\frac{\Gamma}{x^2 + \Gamma^2/4}. \qquad (D.9)$$

The normalization of the distribution to 2π instead of the exact expression is justified in the limit of vanishing width considered here. The Breit-Wigner distribution is a representation of the Dirac delta distribution, therefore

$$\frac{\varepsilon(\text{Im}\Pi_T)}{\pi}\frac{\text{Im}\Pi_T}{\text{Re}^2 D_T^{-1} + \text{Im}^2 \Pi_T} \xrightarrow{\text{Im}\Pi_T \to 0} \delta(\text{Re}D_T^{-1}) \qquad (D.10)$$

in the region $\omega > k$. Consequently, for $\omega > k$, the first term in the curly bracket equals $-\pi\varepsilon(\text{Im}\Pi_T)\delta(\text{Re}D_T^{-1})\text{Re}\Pi_T$. The dispersion relation is valid for $\omega_{T,k}$ so that

$$\delta(\text{Re}D_T^{-1}) = \sum_{\text{zeros } i \text{ of Re}D_T^{-1}} \delta(\omega - \omega_i) / \left|\frac{\partial \text{Re}D_T^{-1}}{\partial \omega}\right|_{\omega_i}$$

$$= \delta(\omega - \omega_{T,k}) / \left|\frac{\partial \text{Re}D_T^{-1}}{\partial \omega}\right|_{\omega_{T,k}}. \qquad (D.11)$$

The derivative of the real part of the inverse transverse gluon propagator with respect to ω is found to be

$$\frac{\partial \text{Re}D_T^{-1}}{\partial \omega} = -2\omega + \frac{m_D^2}{2}\left(\frac{3\omega}{k} - \frac{3\omega - k}{2k^3}\ln\left|\frac{\omega + k}{\omega - k}\right|\right). \qquad (D.12)$$

After substitution of the logarithmic term with $\text{Re}\Pi_T$ and evaluation at the dispersion relation $(\text{Re}\Pi_T(\omega_{T,k}) = \omega_{T,k}^2 - k^2)$ the energy integration of the first term of the curly bracket can be carried out:

$$-\pi \int_k^\infty d\omega \frac{\partial n_B}{\partial T}\varepsilon(\text{Im}\Pi_T)\text{Re}\Pi_T \delta(\omega - \omega_{T,k}) / \left|\frac{\partial \text{Re}D_T^{-1}}{\partial \omega}\right|_{\omega_{T,k}} =$$

$$+\pi \left.\frac{\partial n_B}{\partial T}\right|_{\omega_{T,k}} \omega_{T,k} \frac{(\omega_{T,k}^2 - k^2)^2}{|(\omega_{T,k}^2 - k^2)^2 - m_D^2 \omega_{T,k}^2|}. \qquad (D.13)$$

The energy integral of the second term in the curly bracket vanishes for $\mathrm{Im}\Pi_T \to 0$. Consequently, it does not contribute for $\omega > k$.

For $\omega < k$ the imaginary part of the self-energy Π_T is nonzero except for $\omega = 0$. However, at $\omega = 0$ the real part of the inverse transverse gluon propagator D_T^{-1} is generally nonzero so that no special treatment is necessary. The terms of the curly bracket can be simplified and the final expression for the derivative of the entropy density with respect to μ within $\mathrm{Re}D_T^{-1}$ becomes

$$\left(\frac{\partial s_g}{\partial \mu}\right)_{\mathrm{Re}D_T^{-1}} = +\frac{d_g}{\pi^3 m_D^2}\frac{\partial m_D^2}{\partial \mu}\int_0^\infty dk\, k^2 \qquad (D.14)$$

$$\times \left(\int_0^k d\omega \left[\frac{\partial n_B}{\partial T}\frac{2\mathrm{Re}\Pi_T \mathrm{Im}^3 \Pi_T}{(\mathrm{Re}^2 D_T^{-1} + \mathrm{Im}^2 \Pi_T)^2}\right] - \pi\left.\frac{\partial n_B}{\partial T}\right|_{\omega_{T,k}} \frac{\omega_{T,k}(\omega_{T,k}^2 - k^2)^2}{|(\omega_{T,k}^2 - k^2)^2 - m_D^2 \omega_{T,k}^2|}\right).$$

Analogously we find

$$\left(\frac{\partial s_g}{\partial \mu}\right)_{\mathrm{Re}D_L^{-1}} = +\frac{d_g}{2\pi^3 m_D^2}\frac{\partial m_D^2}{\partial \mu}\int_0^\infty dk\, k^2 \qquad (D.15)$$

$$\times \left(\int_0^k d\omega \left[\frac{\partial n_B}{\partial T}\frac{2\mathrm{Re}\Pi_L \mathrm{Im}^3 \Pi_L}{(\mathrm{Re}^2 D_L^{-1} + \mathrm{Im}^2 \Pi_L)^2}\right] - \pi\left.\frac{\partial n_B}{\partial T}\right|_{\omega_{L,k}} \frac{\omega_{L,k}(\omega_{L,k}^2 - k^2)}{|\omega_{L,k}^2 - k^2 - m_D^2|}\right).$$

The derivatives of the gluon entropy density with respect to μ within the imaginary parts of the self-energies are straightforward. The Dirac delta distributions arising from the the sign functions in the quasiparticle pole contributions $\pi\varepsilon(\mathrm{Im}\Pi_i)\Theta(\mp\mathrm{Re}D_i^{-1})$ vanish due to the prefactor $\mathrm{Im}\Pi_i/m_D^2$ from the chain rule. We are left with

$$\left(\frac{\partial s_g}{\partial \mu}\right)_{\mathrm{Im}\Pi_T} = -\frac{d_g}{\pi^3 m_D^2}\frac{\partial m_D^2}{\partial \mu}\int_0^\infty dk\, k^2 \int_0^\infty d\omega\, \frac{\partial n_B}{\partial T}\frac{2\mathrm{Re}D_T^{-1}\mathrm{Im}^3\Pi_T}{(\mathrm{Re}^2 D_T^{-1} + \mathrm{Im}^2\Pi_T)^2}, \qquad (D.16)$$

$$\left(\frac{\partial s_g}{\partial \mu}\right)_{\mathrm{Im}\Pi_L} = +\frac{d_g}{2\pi^3 m_D^2}\frac{\partial m_D^2}{\partial \mu}\int_0^\infty dk\, k^2 \int_0^\infty d\omega\, \frac{\partial n_B}{\partial T}\frac{2\mathrm{Re}D_L^{-1}\mathrm{Im}^3\Pi_L}{(\mathrm{Re}^2 D_L^{-1} + \mathrm{Im}^2\Pi_L)^2}. \qquad (D.17)$$

Putting things together and substituting $\mathrm{Re}\Pi_T - \mathrm{Re}D_T^{-1} = \omega^2 - k^2$ and $\mathrm{Re}\Pi_L + \mathrm{Re}D_L^{-1} = -k^2$

D.2 HTL quasiparticle model

we have

$$\frac{\partial s_g}{\partial \mu} = \frac{\partial m_D^2}{\partial \mu} \left\{ \frac{d_g}{2\pi^3 m_D^2} \int_0^\infty dk\, k^2 \right. \tag{D.18}$$

$$\times \left(\int_0^k d\omega \left[\frac{\partial n_B}{\partial T} \frac{4(\omega^2 - k^2)\mathrm{Im}^3 \Pi_T}{(\mathrm{Re}^2 D_T^{-1} + \mathrm{Im}^2 \Pi_T)^2} - \frac{\partial n_B}{\partial T} \frac{2k^2 \mathrm{Im}^3 \Pi_L}{(\mathrm{Re}^2 D_L^{-1} + \mathrm{Im}^2 \Pi_L)^2} \right] \right.$$

$$\left. \left. -\pi \frac{\omega_{T,k}(\omega_{T,k}^2 - k^2)^2}{|(\omega_{T,k}^2 - k^2)^2 - m_D^2 \omega_{T,k}^2|} \frac{\partial n_B}{\partial T}\bigg|_{\omega_{T,k}} - \pi \frac{\omega_{L,k}(\omega_{L,k}^2 - k^2)}{|\omega_{L,k}^2 - k^2 - m_D^2|} \frac{\partial n_B}{\partial T}\bigg|_{\omega_{L,k}} \right) \right\}_{(1)}$$

as final expression for the gluons with the derivative of the Debye mass with respect the μ given in Appendix C. The numbered curly bracket is used as abbreviation in the following.

Quarks

The derivative of the quark entropy density with respect to μ is obtained in a similar fashion. Due to the dependence of the Fermi-Dirac distribution function n_F on the chemical potential explicit derivatives emerge. However, these are cancelled in the Maxwell relation by explicit derivatives from the derivative of the particle density with respect to the temperature due to Schwarz's theorem, as for the eQPM models. For convenience starting from Eq. (2.39) only the dependencies of $\mathrm{Re} S_+^{-1}$ and $\mathrm{Im}\Sigma_+$ on μ have to be taken into account:

$$\frac{\partial s_q}{\partial \mu} = \left(\frac{\partial s_q}{\partial \mu}\right)_{\mathrm{Re} S_+^{-1}} + \left(\frac{\partial s_q}{\partial \mu}\right)_{\mathrm{Im}\Sigma_+} + \left(\frac{\partial s_q}{\partial \mu}\right)_{n_F^{(A)}}. \tag{D.19}$$

The third term is the explicit term which is given by Eq. (2.39) with the substitution $\partial n_F^{(A)}/\partial T \to \partial^2 n_F^{(A)}/\partial \mu \partial T$. While the derivative of s_q with respect to μ within $\mathrm{Im}\Sigma_+$ is entirely equivalent to the transverse gluon case, the derivative of s_q with respect to μ in $\mathrm{Re} S_+^{-1}$ differs due to the two dispersion relations $\omega_{\mathrm{TL},k}$ and $\omega_{\mathrm{Pl},k}$ and the different derivative of the real part of the inverse propagator with respect to ω.

For the case[1] $|\omega| > k$ we have

$$\delta(\mathrm{Re} S_+^{-1}) = \delta(\omega - \omega_{\mathrm{TL},k})/\left|\frac{\partial \mathrm{Re} S_+^{-1}}{\partial \omega}\right|_{\omega_{\mathrm{TL},k}} + \delta(\omega - \omega_{\mathrm{Pl},k})/\left|\frac{\partial \mathrm{Re} S_+^{-1}}{\partial \omega}\right|_{\omega_{\mathrm{Pl},k}}, \tag{D.20}$$

where

$$\frac{\partial \mathrm{Re} S_+^{-1}}{\partial \omega} = -1 + \frac{\mathrm{Re}\Sigma_+}{\omega - k} - \frac{2\hat{M}^2}{\omega^2 - k^2}. \tag{D.21}$$

The integral over the first term of the curly bracket times $\pi \varepsilon(\mathrm{Im}\Sigma_+)$ which, up to the different

---

[1] Note that the energy integral limits for both quark contributions are $(-\infty \ldots \infty)$ as opposed to the gluons with $[0 \ldots \infty)$.

propagators/self-energies, is analogous to the transverse gluon case (cf. Eqs. (D.8) and (D.13)) then reads

$$\pi \frac{\omega_{\text{TL},k}^2 - k^2}{2\hat{M}^2}(\omega_{\text{TL},k} - k)\, (+)|_{\omega_{\text{TL},k}} - \pi \frac{\omega_{\text{Pl},k}^2 - k^2}{2\hat{M}^2}(\omega_{\text{Pl},k} + k)\, (+)|_{\omega_{\text{Pl},k}}, \quad (\text{D.22})$$

where

$$(+) := \left(\frac{\partial n_{\text{F}}}{\partial T} + \frac{\partial n_{\text{F}}^A}{\partial T}\right). \quad (\text{D.23})$$

The energy integral of second term of the curly bracket again vanishes for $\text{Im}\Sigma_+ \to 0$ and thus does not contribute for $|\omega| > k$.

For $|\omega| < k$ there is no difference to the transverse gluon expression except for the different propagators and self-energies so that the final quark expression reads

$$\frac{\partial s_q}{\partial \mu} = \frac{\partial \hat{M}^2}{\partial \mu} \Bigg\{ \frac{d_q}{2\pi^3 \hat{M}^2} \int_0^\infty dk\, k^2 \Bigg(\int_{-k}^{k} d\omega \left[(+) \frac{2(\omega - k)\text{Im}^3\Sigma_+}{(\text{Re}^2 S_+^{-1} + \text{Im}^2 \Sigma_+)^2} \right] \quad (\text{D.24})$$

$$-\pi \frac{\omega_{\text{TL},k}^2 - k^2}{2\hat{M}^2}(\omega_{\text{TL},k} - k)\, (+)|_{\omega_{\text{TL},k}} - \pi \frac{\omega_{\text{Pl},k}^2 - k^2}{2\hat{M}^2}(\omega_{\text{Pl},k} + k)\, (+)|_{\omega_{\text{Pl},k}} \Bigg) \Bigg\}_{(I)}$$

with (I) indicating the brackets for quarks. The derivative of $\hat{M}^2$ with respect to μ is given in Appendix C.

The strange quark contribution to the Maxwell relation equals the quark expression at vanishing chemical potential

$$\frac{\partial s_s}{\partial \mu} = \left.\frac{\partial s_q}{\partial \mu}\right|_{\mu=0} \quad (\text{D.25})$$

with curly brackets $\{\}_{(II)} := \{\}_{(I)}|_{\mu=0}$ and $\partial \hat{M}^2/\partial \mu \to \partial \hat{M}_s^2/\partial \mu = (\partial \hat{M}^2/\partial \mu)|_{\mu=0}$.

Particle density

The last step towards the flow equation is the derivative of the particle density with respect to the temperature. The calculation is similar to the the calculation of $\partial s_q/\partial \mu$ with the derivatives $\partial/\partial \mu$ and $\partial/\partial T$ being exchanged. Due to $\partial^2 n_{\text{F}}^{(A)}/\partial \mu \partial T = \partial^2 n_{\text{F}}^{(A)}/\partial T \partial \mu$ it is clear that the explicit derivatives of both cases are equal and cancel within the flow equation. They are, therefore, again omitted. We find

D.2 HTL quasiparticle model

$$\frac{\partial n_q}{\partial T} = \frac{\partial \hat{M}^2}{\partial T} \left\{ \frac{d_q}{2\pi^3 \hat{M}^2} \int_0^\infty dk\, k^2 \left(\int_{-k}^{k} d\omega \left[(\tilde{+}) \frac{2(\omega-k)\mathrm{Im}^3 \Sigma_+}{(\mathrm{Re}^2 S_+^{-1} + \mathrm{Im}^2 \Sigma_+)^2} \right] \right.\right. \tag{D.26}$$

$$\left.\left. -\pi \frac{\omega_{\mathrm{TL},k}^2 - k^2}{2\hat{M}^2}(\omega_{\mathrm{TL},k} - k)\, (\tilde{+})\big|_{\omega_{\mathrm{TL},k}} - \pi \frac{\omega_{\mathrm{Pl},k}^2 - k^2}{2\hat{M}^2}(\omega_{\mathrm{Pl},k} + k)\, (\tilde{+})\big|_{\omega_{\mathrm{Pl},k}} \right) \right\}_{(A)}$$

with

$$(\tilde{+}) := \left(\frac{\partial n_F}{\partial \mu} + \frac{\partial n_F^A}{\partial \mu} \right) \tag{D.27}$$

and $\partial \hat{M}^2/\partial T$ given in Appendix C. For the heavy quark flavor and gluons $(\tilde{+})$, and thus $\partial n_{g,s}/\partial T$, vanishes as a consequence of $\mu_{g,s} = 0$.

The flow equation

Using the results above and the derivatives of the gluon/fermion mass parameters from Appendix C the Maxwell relation assumes the form of a flow equation

$$a_T \frac{\partial G^2}{\partial T} + a_\mu \frac{\partial G^2}{\partial \mu} = b \tag{D.28}$$

as partial differential equation for the effective coupling G^2 with the coefficients[2]

$$\begin{aligned}
a_T &= -\tilde{C}_f \{\}_{(A)}, \\
a_\mu &= 2\tilde{C}_b \{\}_{(1)} + \tilde{C}_f \{\}_{(I)} + \frac{C_f}{8} T^2 \{\}_{(II)}, \\
b &= \frac{C_f}{4} T G^2 \{\}_{(A)} - \frac{N_c N_l}{3\pi^2} \mu G^2 \{\}_{(1)} - \frac{C_f}{4\pi^2} \mu G^2 \{\}_{(I)}.
\end{aligned} \tag{D.29}$$

---

[2] Comparing these coefficients to the results from [Rom04] (Eqs. (B.1) to (B.5)) several differences are noticeable:

1. The expression $(\omega_{\mathrm{T},k}^2 - k^2)$ in the numerator of $\omega_{\mathrm{T},k}(\omega_{\mathrm{T},k}^2 - k^2)^2/|(\omega_{\mathrm{T},k}^2-k^2)^2 - m_D^2 \omega_{\mathrm{T},k}^2|$ within bracket $\{\}_{(1)}$ (Eq. (D.18)) is not squared. This is incorrect, as the dimension of the term thus differs from the terms it is being added to.

2. The terms $(\omega_{i,k}^2 - k^2)/(2\hat{M}^2)$ in $-\pi(\omega_{i,k}^2-k^2)(\omega_{i,k}-k)\,(\tilde{+})\big|_{\omega_{i,k}}/2\hat{M}^2$ in bracket $\{\}_{(A)}$ (Eq. (D.26)) are missing. This is most probably a typographical error, as the neglect of the term leads to an increased a_T and thus to characteristics reaching $T=0$ at very small values of the chemical potential.

3. The coefficient b from Eq. (D.29) and the same coefficient found by Romatschke are related by $b^{\mathrm{Rom}} = -b/G^2$. As a consequence the coupling G^2 from [Rom04] would decrease for $\mu \gtrsim 0$ as opposed to any of the other models. Therefore, we believe this to be incorrect, too.

As a consequence, the results obtained for $N_f = 2$ in [Rom04] are not reproducible.

The pressure Taylor coefficients

The susceptibilities (4.2) of the HTL QPM follow from the pressure (2.44) with partial pressures (2.42) or, more directly, from the net quark density (2.53)

$$
\begin{aligned}
c_2 &= -\frac{d_q T}{2\pi^3} \int_0^\infty dk\, k^2 \int_0^\infty d\omega\, \frac{\partial^2 n_F}{\partial \mu^2} \{\}_{q,+}, \\
c_4 &= -\frac{d_q}{24\pi^3} \left[\left(\int_0^\infty dk\, k^2 \int_0^\infty d\omega\, \frac{\partial^4 n_F}{\partial \mu^4} \{\}_{q,+} \right) - \frac{3\pi^3}{d_q} \frac{\partial^2 \hat{M}^2}{\partial \mu^2} \{\}_{(2A)} \right]_{\mu=0}
\end{aligned}
\quad (\text{D.30})
$$

with the abbreviation $\{\}_{q,+}$ as introduced in Section 2.8 and the abbreviation $\{\}_{(2A)}$ representing $\{\}_{(A)}$ as above with $\partial n_F/\partial \mu \longrightarrow \partial^2 n_F/\partial \mu^2$. Using the short form of the distribution functions as defined in Section 3.3 we have $\partial^4 n_F/\partial \mu^4 = (e^4 - 11e^3 + 11e^2 - e)f^5/T^4$ at $\mu = 0$. As necessary, the odd c_i vanish.

The second derivative of the plasma frequency $\hat{M}^2$ is given in Appendix C with the second derivative of the effective coupling with respect to the chemical potential contained therein following from the coefficients of the flow equation (cf. Eq. (4.5)) and their derivatives. We have

$$
\begin{aligned}
\left.\frac{\partial a_T}{\partial \mu}\right|_{\mu=0} &= -\left(\frac{C_f}{8} T^2 \{\}_{(2A)}\right)_{\mu=0}, \\
\left.\frac{\partial b}{\partial \mu}\right|_{\mu=0} &= G^2 \left(\frac{C_f}{4} T \{\}_{(2A)} - \frac{N_c N_l}{3\pi^2} \{\}_{(1)} - \frac{C_f}{4\pi^2} \{\}_{(I)}\right)_{\mu=0}
\end{aligned}
\quad (\text{D.31})
$$

for the two necessary derivatives.

The mean field pressure

For the integration of the mean field pressure the derivatives of the partial pressures on T, μ and G^2 via the self-energies can be written formally as chain rule (cf. Eq. (2.62)). In practice, it is more prudent to calculate the derivatives directly, so that

$$
\begin{aligned}
\frac{\partial B_g}{\partial [T|\mu|G^2]} &= \left.\frac{\partial m_D^2}{\partial []}\right|_{\text{others constant}} \{\}_{[1]}, \\
\frac{\partial B_q}{\partial [T|\mu|G^2]} &= \left.\frac{\partial \hat{M}^2}{\partial []}\right|_{\text{others constant}} \{\}_{[I]},
\end{aligned}
$$

where the strange quark contribution $\partial B_s/\partial[]$ is given by setting $\mu = 0$ in the light quark result. The abbreviations $\{\}_{[i]}$ correspond to the $\{\}_{(i)}$ via a replacement of the derivatives of the distribution functions in the latter by the distribution functions themselves in the former.

Glossary of abbreviations

CJT	Cornwall-Jackiw-Tomboulis
CBM	compressed baryonic matter
EOS	equation of state
eQPM	effective QPM
eQPM+l	effective QPM with leptons
FAIR	Facility for Antiproton and Ion Research
HIC	heavy-ion collisions
HRG	hadron resonance gas
HTL	hard thermal loop
IR	infrared
LD	Landau damping
LW	Luttinger-Ward
QCD	quantum chromodynamics
QED	quantum electrodynamics
QGP	quark-gluon plasma
QPM	quasiparticle model
SPS	Super Proton Synchrotron
TOV	Tolman-Oppenheimer-Volkoff
UV	ultraviolet

Bibliography

[AB77]   W. D. Arnett, R. L. Bowers, *A Microscopic Interpretation of Neutron Star Structure.* Astrophys. J. Suppl. **33**, 415 (1977), doi:10.1086/190434.

[Alf08] M. G. Alford, A. Schmitt, K. Rajagopal, T. Schäfer, *Color superconductivity in dense quark matter.* Rev. Mod. Phys. **80**, 1455 (2008), doi:10.1103/RevModPhys.80.1455, arXiv:0709.4635.

[All02] C. R. Allton et al., *QCD thermal phase transition in the presence of a small chemical potential.* Phys. Rev. D **66**, 074507 (2002), doi:10.1103/PhysRevD.66.074507, arXiv:hep-lat/0204010.

[All03] C. R. Allton et al., *Equation of state for two flavor QCD at nonzero chemical potential.* Phys. Rev. D **68**, 014507 (2003), doi:10.1103/PhysRevD.68.014507, arXiv:hep-lat/0305007.

[All05] C. R. Allton et al., *Thermodynamics of two flavor QCD to sixth order in quark chemical potential.* Phys. Rev. D **71**, 054508 (2005), doi:10.1103/PhysRevD.71.054508, arXiv:hep-lat/0501030.

[Ana79] J. D. Anand, P. P. Bhattacharjee, S. N. Biswas, *Possible existence of quark stars.* J. Phys. A **12**, L347 (1979), doi:10.1088/0305-4470/12/12/006.

[And10a] J. O. Andersen, M. Strickland, N. Su, *Three-loop HTL gluon thermodynamics at intermediate coupling.* JHEP **1008**, 113 (2010), doi:10.1007/JHEP08(2010)113, arXiv:1005.1603.

[And10b] J. O. Andersen, L. E. Leganger, M. Strickland, N. Su, *NNLO hard-thermal-loop thermodynamics for QCD* (2010), arXiv:1009.4644.

[Ans03] M. Ansorg, A. Kleinwächter, R. Meinel, *Highly accurate calculation of rotating neutron stars.* Astron.Astrophys. **405**, 711 (2003), doi:10.1051/0004-6361:20030618, arXiv:astro-ph/0301173.

[Arr02] A. Arrizabalaga, J. Smit, *Gauge-fixing dependence of Φ-derivable approximations.* Phys. Rev. D **66**, 065014 (2002), doi:10.1103/PhysRevD.66.065014, arXiv:hep-ph/0207044.

[AS02]   J. O. Andersen, M. Strickland, *Equation of state for dense QCD and quark stars.* Phys. Rev. D **66**, 105001 (2002), doi:10.1103/PhysRevD.66.105001, arXiv:hep-ph/0206196.

[Bar66]   J. Bardeen, G. Baym, D. Pines, *Interactions Between He^3 Atoms in Dilute Solutions of He^3 in Superfluid He^4.* Phys. Rev. Lett. **17**, 372 (1966), doi:10.1103/PhysRevLett.17.372.

[Bar89]   H. W. Barz, B. L. Friman, J. Knoll, H. Schulz, *Nuclear-matter—quark-matter phase diagram with strangeness.* Phys. Rev. D **40**, 157 (1989), doi:10.1103/PhysRevD.40.157.

[Bay62]   G. Baym, *Self-Consistent Approximations in Many-Body Systems.* Phys. Rev. **127**, 1391 (1962), doi:10.1103/PhysRev.127.1391.

[Bay76]   G. Baym, S. A. Chin, *Can a neutron star be a giant MIT bag?* Phys. Lett. B **62**, 241 (1976), doi:10.1016/0370-2693(76)90517-7.

[Baz09]   A. Bazavov et al., *Equation of state and QCD transition at finite temperature.* Phys. Rev. D **80**, 014504 (2009), doi:10.1103/PhysRevD.80.014504, arXiv:0903.4379.

[Bec05]   W. C. Beckmann, *Self-consistent Calculations of Hadron Properties at Non-zero Temperature.* Ph.D. thesis, University of Frankfurt a. M. (2005), http://deposit.ddb.de/cgi-bin/dokserv?idn=979509920.

[Ber04]   J. Berges, *n-Particle irreducible effective action techniques for gauge theories.* Phys. Rev. D **70**, 105010 (2004), doi:10.1103/PhysRevD.70.105010, arXiv:hep-ph/0401172.

[Bha10]   A. Bhattacharyya, I. N. Mishustin, W. Greiner, *Deconfinement phase transition in compact stars: Maxwell versus Gibbs construction of the mixed phase.* J. Phys. G **37**, 025201 (2010), doi:10.1088/0954-3899/37/2/025201, arXiv:0905.0352.

[BIR01]   J.-P. Blaizot, E. Iancu, A. Rebhan, *Approximately self-consistent resummations for the thermodynamics of the quark-gluon plasma: Entropy and density.* Phys. Rev. D **63**, 065003 (2001), doi:10.1103/PhysRevD.63.065003, arXiv:hep-ph/0005003.

[BK61]   G. Baym, L. P. Kadanoff, *Conservation Laws and Correlation Functions.* Phys. Rev. **124**, 287 (1961), doi:10.1103/PhysRev.124.287.

[BKS06]   M. Bluhm, B. Kämpfer, R. Schulze, D. Seipt, *Isentropic Equation of State of Two-Flavour QCD in a Quasi-Particle Model.* Acta Phys. Hung. A **27**, 397 (2006), doi:10.1556/APH.27.2006.4.2, arXiv:hep-ph/0608052.

[BKS07a] M. Bluhm, B. Kämpfer, R. Schulze, D. Seipt, *Quasi-particle description of strongly interacting matter: Towards a foundation.* Eur. Phys. J. C **49**, 205 (2007), doi:10.1140/epjc/s10052-006-0056-y, arXiv:hep-ph/0608053.

[BKS07b] M. Bluhm, B. Kämpfer, R. Schulze, D. Seipt, U. Heinz, *A family of equations of state based on lattice QCD: Impact on flow in ultrarelativistic heavy-ion collisions.* Phys. Rev. C **76**, 034901 (2007), arXiv:0705.0397.

[Bla00] D. Blaschke, T. Klähn, D. N. Voskresensky, *Diquark Condensates and Compact Star Cooling.* Astrophys. J. **533**, 406 (2000), doi:10.1086/308664.

[Blu04] M. Bluhm, *On the equation of state of strongly interacting matter - A quasi-particle description.* Diploma thesis, Technical University Dresden (2004).

[Blu04a] M. Bluhm, B. Kämpfer, G. Soff, *A unique parameterization of the QCD equation of state below and above $T(c)$* (2004), arXiv:hep-ph/0402252.

[Blu05a] M. Bluhm, B. Kämpfer, G. Soff, *The QCD equation of state near T_c within a quasi-particle model.* Phys. Lett. B **620**, 131 (2005), doi:10.1016/j.physletb.2005.05.083, arXiv:hep-ph/0411106.

[Blu05b] M. Bluhm, B. Kämpfer, *Quasi-particle perspective on QCD matter and critical end point effects* pp. 124–128 (2005), arXiv:hep-ph/0511015.

[Blu06] M. Bluhm, B. Kämpfer, *Quasi-particle perspective on critical end-point.* PoS **CPOD2006**, 004 (2006), arXiv:hep-ph/0611083.

[Blu08a] M. Bluhm, B. Kämpfer, *Quasiparticle model of quark-gluon plasma at imaginary chemical potential.* Phys. Rev. D **77**, 034004 (2008), doi:10.1103/PhysRevD.77.034004, arXiv:0711.0590.

[Blu08b] M. Bluhm, B. Kämpfer, *Flavor diagonal and off-diagonal susceptibilities in a quasiparticle model of the quark-gluon plasma.* Phys. Rev. D **77**, 114016 (2008), doi:10.1103/PhysRevD.77.114016, arXiv:0801.4147.

[Blu08c] M. Bluhm, *QCD equation of state of hot deconfined matter at finite baryon density - A quasiparticle perspective.* Ph.D. thesis, Technical University Dresden (2008), http://www.fzd.de/db/Cms?pOid=27947.

[Bom04] I. Bombaci, I. Parenti, I. Vidaña, *Quark Deconfinement and Implications for the Radius and the Limiting Mass of Compact Stars.* ApJ **614**, 314 (2004), doi:10.1086/423658, arXiv:astro-ph/0402404.

[Bor10a] S. Borsanyi et al. (Wuppertal-Budapest Collaboration), *Is there still any T_c mystery in lattice QCD? Results with physical masses in the continuum limit III* (2010), arXiv:1005.3508.

[Bor10b] S. Borsanyi et al. (Wuppertal-Budapest Collaboration), *The QCD equation of state with dynamical quarks*. JHEP (2010), accepted for publication, arXiv:1007.2580.

[BP90a] E. Braaten, R. D. Pisarski, *Resummation and gauge invariance of the gluon damping rate in hot QCD*. Phys. Rev. Lett. **64**, 1338 (1990), doi:10.1103/PhysRevLett.64.1338.

[BP90b] E. Braaten, R. D. Pisarski, *Soft amplitudes in hot gauge theories: A general analysis*. Nucl. Phys. B **337**, 569 (1990), doi:10.1016/0550-3213(90)90508-B.

[Bro92] L. S. Brown, *Quantum Field Theory* (Cambridge University Press, 1992), ISBN 0521469465.

[Car04] M. E. Carrington, *The 4PI effective action for ϕ^4 theory*. Eur. Phys. J. C **35**, 383 (2004), doi:10.1140/epjc/s2004-01849-6, arXiv:hep-ph/0401123.

[CC74] V. Canuto, S. M. Chitre, *Crystallization of dense neutron matter*. Phys. Rev. D **9**, 1587 (1974), doi:10.1103/PhysRevD.9.1587.

[CH98] S. Chiku, T. Hatsuda, *Optimized perturbation theory at finite temperature*. Phys. Rev. D **58**, 076001 (1998), doi:10.1103/PhysRevD.58.076001.

[Cha08] C. J. Chammasheril, P. C. Vinodkumar, *Sensitivity of quark matter EoS parameters and quark star properties*. J. Phys.: Conf. Ser. **110**, 062004 (2008), http://stacks.iop.org/1742-6596/110/i=6/a=062004.

[Che07] M. Cheng et al. (RBC-Bielefeld Collaboration), *The QCD Equation of State with almost Physical Quark Masses*. Phys. Rev. D **77**, 014511 (2008), doi:10.1103/PhysRevD.77.014511, arXiv:0710.0354.

[CJT74] J. M. Cornwall, R. Jackiw, E. Tomboulis, *Effective action for composite operators*. Phys. Rev. D **10**, 2428 (1974), doi:10.1103/PhysRevD.10.2428.

[Cle06] J. Cleymans, H. Oeschler, K. Redlich, S. Wheaton, *Comparison of chemical freeze-out criteria in heavy-ion collisions*. Phys. Rev. C **73**, 034905 (2006), doi:10.1103/PhysRevC.73.034905, arXiv:hep-ph/0511094.

[CP75] G. M. Carneiro, C. J. Pethick, *Specific heat of a normal Fermi liquid. II. Microscopic approach*. Phys. Rev. B **11**, 1106 (1975), doi:10.1103/PhysRevB.11.1106.

[DMB69] R. Dashen, S.-k. Ma, H. J. Bernstein, *S-Matrix Formulation of Statistical Mechanics.* Phys. Rev. **187**, 345 (1969), doi:10.1103/PhysRev.187.345.

[dDM64] C. De Dominicis, P. C. Martin, *Stationary Entropy Principle and Renormalization in Normal and Superfluid Systems. I. Algebraic Formulation.* J. Math. Phys. **5**, 14 (1964), doi:10.1063/1.1704062.

[DEl04] M. D'Elia, M.-P. Lombardo, *QCD thermodynamics from an imaginary μ_B: Results on the four flavor lattice model.* Phys. Rev. D **70**, 074509 (2004), doi:10.1103/PhysRevD.70.074509, arXiv:hep-lat/0406012.

[DEl07] M. D'Elia, F. Di Renzo, M. P. Lombardo, *Strongly interacting quark-gluon plasma, and the critical behavior of QCD at imaginary μ.* Phys. Rev. D **76**, 114509 (2007), doi:10.1103/PhysRevD.76.114509, arXiv:0705.3814.

[DeT10] C. DeTar et al., *QCD thermodynamics with nonzero chemical potential at $N_t = 6$ and effects from heavy quarks.* Phys. Rev. D **81**, 114504 (2010), doi:10.1103/PhysRevD.81.114504, arXiv:1003.5682.

[DJ67] H. D. Dahmen, G. Jona-Lasinio, *Variational Formulation of Quantum Field Theory. - I.* Nuovo Cimento **A52**, 807 (1967).

[DJ69] H. D. Dahmen, G. Jona-Lasinio, *Variational Formulation of Quantum Field Theory. - II. A study of a Functional Derivative Equation Related to the gA^3 Theory.* Nuovo Cimento **A62**, 889 (1969).

[DJT72] H. D. Dahmen, G. Jona-Lasinio, J. Tarski, *The method of characteristics for functional-derivative equations.* Nuovo Cimento **A10**, 513 (1972).

[Fec78] W. B. Fechner, P. C. Joss, *Quark stars with 'realistic' equations of state.* Nature **274**, 347 (1978), doi:10.1038/274347a0.

[Fra01] E. S. Fraga, R. D. Pisarski, J. Schaffner-Bielich, *Small, dense quark stars from perturbative QCD.* Phys. Rev. D **63**, 121702(R) (2001), doi:10.1103/PhysRevD.63.121702, arXiv:hep-ph/0101143.

[Fra02] E. S. Fraga, R. D. Pisarski, J. Schaffner-Bielich, *New class of compact stars at high density.* Nucl. Phys. A **702**, 217 (2002), doi:10.1016/S0375-9474(02)00709-1, arXiv:nucl-th/0110077.

[Fre77] B. A. Freedman, L. D. McLerran, *Fermions and gauge vector mesons at finite temperature and density. III. The ground-state energy of a relativistic quark gas.* Phys. Rev. D **16**, 1169 (1977), doi:10.1103/PhysRevD.16.1169.

[Fre78]   B. A. Freedman, L. D. McLerran, *Quark star phenomenology.* Phys. Rev. D **17**, 1109 (1978), doi:10.1103/PhysRevD.17.1109.

[FP81]   B. Friedman, V. R. Pandharipande, *Hot and cold, nuclear and neutron matter.* Nucl. Phys. A **361**, 502 (1981), doi:10.1016/0375-9474(81)90649-7.

[FW71]   P. Fulde, H. Wagner, *Low-Temperature Specific Heat and Thermal Conductivity of Noncrystalline Solids.* Phys. Rev. Lett. **27**, 1280 (1971), doi:10.1103/PhysRevLett.27.1280.

[Gar09]   F. Gardim, F. Steffens, *Thermodynamics of quasi-particles at finite chemical potential.* Nucl. Phys. A **825**, 222 (2009), doi:10.1016/j.nuclphysa.2009.05.001, arXiv:0905.0667.

[Gav05a] R. V. Gavai, S. Gupta, *On the critical end point of QCD.* Phys. Rev. D **71**, 114014 (2005), doi:10.1103/PhysRevD.71.114014, arXiv:hep-lat/0412035.

[Gav05b] R. V. Gavai, S. Gupta, *Simple patterns for nonlinear susceptibilities near T_c.* Phys. Rev. D **72**, 054006 (2005), doi:10.1103/PhysRevD.72.054006, arXiv:hep-lat/0507023.

[Ger68]   U. H. Gerlach, *Equation of State at Supranuclear Densities and the Existence of a Third Family of Superdense Stars.* Phys. Rev. **172**, 1325 (1968), doi:10.1103/PhysRev.172.1325.

[Gle97]   N. K. Glendenning, S. Pei, F. Weber, *Signal of Quark Deconfinement in the Timing Structure of Pulsar Spin-Down.* Phys. Rev. Lett. **79**, 1603 (1997), doi:10.1103/PhysRevLett.79.1603, arXiv:astro-ph/9705235.

[Gle00]   N. K. Glendenning, *Compact stars: nuclear physics, particle physics, and general relativity* (Springer, 2000), ISBN 0387989773.

[Gre64]   O. W. Greenberg, *Spin and Unitary-Spin Independence in a Paraquark Model of Baryons and Mesons.* Phys. Rev. Lett. **13**, 598 (1964), doi:10.1103/PhysRevLett.13.598.

[GTP11]   W. Greiner, S. Schramm, E. Stein, *Quantum Chromodynamics*, vol. 11 of *Classical Theoretical Physics* (Springer, 2002), 2nd edn., ISBN 3540666109.

[GW65]   W. Götze, H. Wagner, *On the T^3-law for the specific heat of bose liquids.* Physica **31**, 475 (1965), doi:10.1016/0031-8914(65)90074-1.

[Hae07]   P. Haensel, A. Y. Potekhin, D. G. Yakovlev, *Neutron stars: Equation of state and structure* (Springer, 2006), 1st edn., ISBN 0387335439.

[Hag80]   R. Hagedorn, J. Rafelski, *Hot hadronic matter and nuclear collisions*. Phys. Lett. B **97**, 136 (1980), doi:10.1016/0370-2693(80)90566-3.

[Han04]   M. Hanauske, *Eigenschaften von kompakten Sternen in QCD motivierten Modellen*. Ph.D. thesis, Johann Wolfgang Goethe-Universität Frankfurt/M. (2004), http://publikationen.ub.uni-frankfurt.de/volltexte/2005/587/index.html.

[Har75]   J. B. Hartle, R. F. Sawyer, D. J. Scalapino, *Pion condensed matter at high densities - Equation of state and stellar models*. Astrophys. J. **199**, 471 (1975), doi:10.1086/153713.

[Hil10]   T. Hilger, R. Schulze, B. Kämpfer, *QCD sum rules for D mesons in dense and hot nuclear matter*. J. Phys. G **37**, 094054 (2010), doi:10.1088/0954-3899/37/9/094054, arXiv:1001.0522.

[Hin83]   A. C. Hindmarsh, *ODEPACK, A Systematized Collection of ODE Solvers*. in *Scientific Computing*, edited by R. S. Stepleman et al., vol. 1 of *IMACS Transactions on Scientific Computation*, pp. 55–64 (North-Holland, Amsterdam, 1983), https://computation.llnl.gov/casc/nsde/pubs/u88007.pdf.

[Huo10]   P. Huovinen, P. Petreczky, *QCD equation of state and hadron resonance gas*. Nucl. Phys. A **837**, 26 (2010), doi:10.1016/j.nuclphysa.2010.02.015, arXiv:0912.2541.

[Ito70]   N. Itoh, *Hydrostatic Equilibrium of Hypothetical Quark Stars*. Prog. Theor. Phys. **44**, 291 (1970), doi:10.1143/PTP.44.291.

[Iva05]   Y. B. Ivanov, A. S. Khvorostukhin, E. E. Kolomeitsev, V. V. Skokov, V. D. Toneev, D. N. Voskresensky, *Lattice QCD constraints on hybrid and quark stars*. Phys. Rev. C **72**, 025804 (2005), doi:10.1103/PhysRevC.72.025804, arXiv:astro-ph/0501254v2.

[Iva06]   Y. B. Ivanov, V. N. Russkikh, V. D. Toneev, *Relativistic heavy-ion collisions within three-fluid hydrodynamics: Hadronic scenario*. Phys. Rev. C **73**, 044904 (2006), doi:10.1103/PhysRevC.73.044904, arXiv:nucl-th/0503088.

[Jon64]   G. Jona-Lasinio, *Relativistic Field Theories with Symmetry-Breaking Solutions*. Nuovo Cimento **34**, 1790 (1964).

[Kac10]   O. Kaczmarek et al., *The phase boundary for the chiral transition in (2+1)-flavor QCD at small values of the chemical potential* (2010), arXiv:1011.3130.

[Kal84]   O. Kalashnikov, *QCD at finite temperature*. Fortsch. Phys. **32**, 525 (1984), doi:10.1002/prop.19840321002.

[Kam81a]   B. Kämpfer, *On the possibility of stable quark and pion-condensed stars*. J. Phys. A **14**, L471 (1981), doi:10.1088/0305-4470/14/11/009.

[Kam81b] B. Kämpfer, *On stabilizing effects of relativity in cold spheric stars with a phase transition in the interior*. Phys. Lett. B **101**, 366 (1981), doi:10.1016/0370-2693(81)90065-4.

[Kam83] B. Kämpfer, *Phase transitions in dense nuclear matter and consequences for neutron stars*. J. Phys. G **9**, 1487 (1983), doi:10.1088/0305-4616/9/12/009.

[Kam85] B. Kämpfer, *Phase transitions in dense nuclear matter and explosive neutron star phenomena*. Phys. Lett. B **153**, 121 (1985), doi:10.1016/0370-2693(85)91410-8.

[Kam94] B. Kämpfer, B. Lukács, G. Paál, *Cosmic Phase Transitions* (Teubner Verlagsgesellschaft Leipzig, 1994), ISBN 3815430194.

[Kap89] J. I. Kapusta, *Finite-Temperature Field Theory* (Cambridge University Press, 1989), ISBN 0521449456.

[Kar03] F. Karsch, K. Redlich, A. Tawfik, *Thermodynamics at non-zero baryon number density: A comparison of lattice and hadron resonance gas model calculations*. Phys. Lett. B **571**, 67 (2003), doi:10.1016/j.physletb.2003.08.001, arXiv:hep-ph/0306208.

[Kar07] F. Karsch (RBC-Bielefeld Collaboration), *Transition temperature in QCD with physical light and strange quark masses*. J. Phys. G **34**, S627 (2007), doi:10.1088/0954-3899/34/8/S59, arXiv:hep-ph/0701210.

[KBS06] B. Kämpfer, M. Bluhm, R. Schulze, D. Seipt, U. Heinz, *QCD matter within a quasi-particle model and the critical end point*. Nucl. Phys. A **774**, 757 (2006), doi:10.1016/j.nuclphysa.2006.06.131, arXiv:hep-ph/0509146.

[Kei76] B. D. Keister, L. S. Kisslinger, *Free-quark phases in dense stars*. Phys. Lett. B **64**, 117 (1976), doi:10.1016/0370-2693(76)90370-1.

[Kle82] H. Kleinert, *Higher effective actions for bose systems*. Fortsch. Phys. **30**, 187 (1982), doi:10.1002/prop.19820300402.

[Kli81] V. V. Klimov, *Spectrum of Elementary Fermi Excitations in Quark Gluon Plasma*. Sov. J. Nucl. Phys. (engl. translation) **33**, 934 (1981).

[Kno00] R. Knobel, *An introduction to the mathematical theory of waves* (American Mathematical Society, 2000), ISBN 0821820397.

[KSS06] R. Klanner, T. Schörner-Sadenius, *Verstehen wir die starke Kraft?* Phys. J. **5**, 41 (2006), http://www.pro-physik.de/Phy/pdfstart.do?recordid=23604.

[Kur09] A. Kurkela, P. Romatschke, A. Vuorinen, *Cold Quark Matter* (2009), arXiv:0912.1856.

[Lat00]   J. M. Lattimer, M. Prakash, *Nuclear matter and its role in supernovae, neutron stars and compact object binary mergers*. Phys. Rept. **333-334**, 121 (2000), doi:DOI: 10.1016/S0370-1573(00)00019-3, arXiv:astro-ph/0002203.

[LeB96]   M. L. Bellac, *Thermal Field Theory* (Cambridge University Press, 1996), ISBN 0521654777.

[LFK98]   C. Lobban, J. L. Finney, W. F. Kuhs, *The structure of a new phase of ice*. Nature **391**, 268 (1998), doi:10.1038/34622.

[LL06]   L. D. Landau, E. M. Lifshitz, *Fluid mechanics*, vol. 6 of *Course of Theoretical Physics* (Pergamon, 1987), 2nd edn.

[LRP93]   C. P. Lorenz, D. G. Ravenhall, C. J. Pethick, *Neutron star crusts*. Phys. Rev. Lett. **70**, 379 (1993), doi:10.1103/PhysRevLett.70.379.

[LR03]   J. Letessier, J. Rafelski, *QCD equations of state and the quark-gluon plasma liquid model*. Phys. Rev. C **67**, 031902 (2003), doi:10.1103/PhysRevC.67.031902, arXiv:hep-ph/0301099.

[LW60]   J. M. Luttinger, J. C. Ward, *Ground-State Energy of a Many-Fermion System. II*. Phys. Rev. **118**, 1417 (1960), doi:10.1103/PhysRev.118.1417.

[LY60b]   T. D. Lee, C. N. Yang, *Many-Body Problem in Quantum Statistical Mechanics. IV. Formulation in Terms of Average Occupation Number in Momentum Space*. Phys. Rev. **117**, 22 (1960), doi:10.1103/PhysRev.117.22.

[Maj10]   A. Majumder, B. Müller, *Hadron Mass Spectrum from Lattice QCD* (2010), arXiv:1008.1747.

[MP07]   L. D. McLerran, R. D. Pisarski, *Phases of dense quarks at large Nc*. Nucl. Phys. A **796**, 83 (2007), doi:DOI: 10.1016/j.nuclphysa.2007.08.013, arXiv:0706.2191.

[MTW73]   C. W. Misner, K. S. Thorne, J. A. Wheeler, *Gravitation* (W. H. Freeman), 1st edn., ISBN 0716703440.

[Mor08]   K. Morita, S. H. Lee, *Mass Shift and Width Broadening of J/ψ in Hot Gluonic Plasma from QCD Sum Rules*. Phys. Rev. Lett. **100**, 022301 (2008), doi:10.1103/PhysRevLett.100.022301, arXiv:0704.2021.

[Nam74]   Y. Nambu, M. Y. Han, *Three triplets, paraquarks, and "colored" quarks*. Phys. Rev. D **10**, 674 (1974), doi:10.1103/PhysRevD.10.674.

[NC75]   R. E. Norton, J. M. Cornwall, *On the formalism of relativistic many body theory*. Ann. Phys. **91**, 106 (1975), doi:10.1016/0003-4916(75)90281-X.

[Non05] C. Nonaka, M. Asakawa, *Hydrodynamical evolution near the QCD critical end point.* Phys. Rev. C **71**, 044904 (2005), doi:10.1103/PhysRevC.71.044904, arXiv:nucl-th/0410078.

[ONC99] T. M. O'Neil, F. V. Coroniti, *The collisionless nature of high-temperature plasmas.* Rev. Mod. Phys. **71**, S404 (1999), doi:10.1103/RevModPhys.71.S404.

[OV39] J. R. Oppenheimer, G. M. Volkoff, *On Massive Neutron Cores.* Phys. Rev. **55**, 374 (1939), doi:10.1103/PhysRev.55.374.

[PDG06] W.-M. Yao et al. (Particle Data Group), *Review of Particle Physics.* J. Phys. G **33**, 1 (2006), doi:10.1088/0954-3899/33/1/001, http://pdg.lbl.gov/2006/pdg_2006.html.

[PDG08] C. Amsler et al. (Particle Data Group), *Review of Particle Physics.* Phys. Lett. B **667**, 1 (2008), doi:10.1016/j.physletb.2008.07.018, http://pdg.lbl.gov/2008/pdg_2008.html.

[PDG10] K. Nakamura (Particle Data Group), *Review of Particle Physics.* J. Phys. G **37**, 075021 (2010), doi:10.1088/0954-3899/37/7A/075021, http://pdg.lbl.gov.

[Pei00] A. Peikert, *QCD thermodynamics with 2+1 flavours in lattice simulations.* Ph.D. thesis, University of Bielefeld (2000), http://www.physik.uni-bielefeld.de/theory/e6/publicphd.html.

[Pes94] A. Peshier, B. Kämpfer, O. P. Pavlenko, G. Soff, *An Effective model of the quark - gluon plasma with thermal parton masses.* Phys. Lett. B **337**, 235 (1994), doi:10.1016/0370-2693(94)90969-5.

[Pes96] A. Peshier, B. Kämpfer, O. P. Pavlenko, G. Soff, *Massive quasiparticle model of the SU(3) gluon plasma.* Phys. Rev. D **54**, 2399 (1996), doi:10.1103/PhysRevD.54.2399.

[Pes98] A. Peshier, *Zur Zustandsgleichung heißer stark wechselwirkender Materie - konsistente Beschreibung stark gekoppelter Quantensysteme.* Ph.D. thesis, Technical University Dresden (1998), http://www.fzd.de/publications/001361/1361.pdf.

[Pes00] A. Peshier, B. Kämpfer, G. Soff, *Equation of state of deconfined matter at finite chemical potential in a quasiparticle description.* Phys. Rev. C **61**, 045203 (2000), doi:10.1103/PhysRevC.61.045203, arXiv:hep-ph/9911474.

[Pes01a] A. Peshier, *Hard thermal loop resummation of the thermodynamic potential.* Phys. Rev. D **63**, 105004 (2001), doi:10.1103/PhysRevD.63.105004, arXiv:hep-ph/0011250.

[Pes01b] A. Peshier, B. Kämpfer, G. Soff, *Stringent limits on quark star masses due to the chiral transition temperature* (2001), arXiv:hep-ph/0106090.

[Pes02] A. Peshier, B. Kämpfer, G. Soff, *From QCD lattice calculations to the equation of state of quark matter*. Phys. Rev. D **66**, 094003 (2002), doi:10.1103/PhysRevD.66.094003, arXiv:hep-ph/0206229.

[Pes03] A. Peshier, B. Kämpfer, G. Soff, *The QCD equation of state and quark star properties*. in *Erevan 2003: Superdense QCD matter and compact stars*, edited by D. Blaschke, D. Sedrakian, pp. 135–146 (Springer, 2006), arXiv:hep-ph/0312080.

[Pes04] A. Peshier, *Hard gluon damping in hot QCD*. Phys. Rev. D **70**, 034016 (2004), doi:10.1103/PhysRevD.70.034016, arXiv:hep-ph/0403225.

[Pes05] A. Peshier, *Hard parton damping in hot QCD*. J. Phys. G **31**, 371 (2005), doi:10.1088/0954-3899/31/4/046, arXiv:hep-ph/0409270.

[Pis89b] R. D. Pisarski, *Renormalized fermion propagator in hot gauge theories*. Nucl. Phys. A **498**, 423 (1989), doi:10.1016/0375-9474(89)90620-9.

[Pos10] S. Postnikov, M. Prakash, J. M. Lattimer, *Tidal Love numbers of neutron and self-bound quark stars*. Phys. Rev. D **82**, 024016 (2010), doi:10.1103/PhysRevD.82.024016, arXiv:1004.5098.

[PS95] M. E. Peskin, D. V. Schroeder, *An Introduction to Quantum Field Theory* (Perseus Books, 1995), ISBN 0201503972.

[Raj01] K. Rajagopal, F. Wilczek, *Enforced Electrical Neutrality of the Color-Flavor Locked Phase*. Phys. Rev. Lett. **86**, 3492 (2001), doi:10.1103/PhysRevLett.86.3492, arXiv:hep-ph/0012039.

[Rap01] R. Rapp, *Signatures of thermal dilepton radiation at ultrarelativistic energies*. Phys. Rev. C **63**, 054907 (2001), doi:10.1103/PhysRevC.63.054907, arXiv:hep-ph/0010101.

[Ris91] D. H. Rischke, M. I. Gorenstein, H. Stöcker, W. Greiner, *Excluded volume effect for the nuclear matter equation of state*. Z. Phys. C **51**, 485 (1991), doi:10.1007/BF01548574.

[Ris01] D. H. Rischke, D. T. Son, M. A. Stephanov, *Asymptotic Deconfinement in High-Density QCD*. Phys. Rev. Lett. **87**, 062001 (2001), doi:10.1103/PhysRevLett.87.062001, arXiv:hep-ph/0011379.

[Ris03] D. H. Rischke, *The quark-gluon plasma in equilibrium*. Prog. Part. Nucl. Phys. **52**, 197 (2004), doi:10.1016/j.ppnp.2003.09.002, arXiv:nucl-th/0305030.

[Riv88]   R. J. Rivers, *Path integral methods in quantum field theory* (Cambridge University Press, 1988), 1st edn., ISBN 0521368707.

[Roe05]   D. Röder, *Selfconsistent calculations of mesonic properties at nonzero temperature*. Ph.D. thesis, University of Frankfurt a. M. (2005), arXiv:hep-ph/0601146, http://deposit.ddb.de/cgi-bin/dokserv?idn=978020219.

[Rom04]   P. Romatschke, *Quasiparticle description of the hot and dense quark-gluon plasma*. Ph.D. thesis, Technical University Vienna (2004), arXiv:hep-ph/0312152.

[RW00]   K. Rajagopal, F. Wilczek, *The Condensed Matter Physics of QCD*. in *At the Frontier of Particle Physics: Handbook of QCD*, edited by M. Shifman, vol. 3, chap. 35 (World Scientific, 2000), ISBN 9812380280, arXiv:hep-ph/0011333.

[ScB02]   J. Schaffner-Bielich, M. Hanauske, H. Stöcker, W. Greiner, *Phase Transition to Hyperon Matter in Neutron Stars*. Phys. Rev. Lett. **89**, 171101 (2002), doi:10.1103/PhysRevLett.89.171101.

[Sch07]   R. Schulze, *Quasiparticle description of QCD thermodynamics: effects of finite widths, Landau damping and collective excitations*. Diploma thesis, Technical University Dresden (2007), http://www.fzd.de/db/Cms?pOid=29755.

[Sch08a]   R. Schulze, M. Bluhm, B. Kämpfer, *Plasmons, plasminos and Landau damping in a quasiparticle model of the quark-gluon plasma*. Eur. Phys. J. ST **155**, 177 (2008), doi:10.1140/epjst/e2008-00600-8, arXiv:0709.2262.

[Sch08b]   R. Schulze, M. Bluhm, B. Kämpfer, *Equation of state for strongly interacting matter: collective effects, Landau damping and predictions for LHC*. in *XLVI International Winter Meeting on Nuclear Physics, Bormio (Italy)*, edited by I. Iori, A. Tarantola, p. 63 (University of Milan, 2008), arXiv:0803.1571.

[Sch09]   R. Schulze, B. Kämpfer, *Equation of state for QCD matter in a quasiparticle model*. Prog. Part. Nucl. Phys. **62**, 386 (2009), doi:10.1016/j.ppnp.2008.12.017, arXiv:0811.0274.

[Sch10]   R. Schulze, B. Kämpfer, *Cold quark stars from hot lattice QCD* (2009), arXiv:0912.2827.

[Sei07]   D. Seipt, *Quark Mass Dependence of One-Loop Self-Energies in Hot QCD*. Diploma thesis, Technical University Dresden (2007), http://www.fzd.de/db/Cms?pOid=29753.

[Sei09]   D. Seipt, M. Bluhm, B. Kämpfer, *Quark mass dependence of thermal excitations in QCD in one-loop approximation*. J. Phys. G **36**, 045003 (2009), doi:10.1088/0954-3899/36/4/045003, arXiv:0810.3803.

[SLS99]   K. Schertler, S. Leupold, J. Schaffner-Bielich, *Neutron stars and quark phases in the Nambu-Jona-Lasinio model.* Phys. Rev. C **60**, 025801 (1999), doi:10.1103/PhysRevC.60.025801, arXiv:astro-ph/9901152.

[Sta10]   P. Staszel (CBM Collaboration), *CBM experiment at FAIR.* Acta Phys.Polon. B **41**, 341 (2010), http://th-www.if.uj.edu.pl/acta/vol41/abs/v41p0341.htm.

[Ste09]   J. Steinheimer, S. Schramm, H. Stöcker, *An Effective chiral Hadron-Quark Equation of State Part I: Zero baryochemical potential* (2009), arXiv:0909.4421.

[Str10]   M. Strickland, J. O. Andersen, L. E. Leganger, N. Su, *Hard-thermal-loop QCD Thermodynamics* (2010), arXiv:1011.0416.

[SW99]   T. Schäfer, F. Wilczek, *Continuity of Quark and Hadron Matter.* Phys. Rev. Lett. **82**, 3956 (1999), doi:10.1103/PhysRevLett.82.3956, arXiv:hep-ph/9811473.

[TBM01]   I. N. Bronstein, K. A. Semendjajew, *Taschenbuch der Mathematik*, vol. 1 (Teubner Verlagsgesellschaft Leipzig, 1996).

[TL97]   B. T. Tsurutani, G. S. Lakhina, *Some basic concepts of wave-particle interactions in collisionless plasmas.* Rev. Geo. **35**, 491 (1997), doi:10.1029/97RG02200.

[Tol34]   R. C. Tolman, *Effect of Inhomogeneity on Cosmological Models.* Proc. Natl. Acad. Sci. USA **20**, 169 (1934), doi:10.1073/pnas.20.3.169.

[Tol39]   R. C. Tolman, *Static Solutions of Einstein's Field Equations for Spheres of Fluid.* Phys. Rev. **55**, 364 (1939), doi:10.1103/PhysRev.55.364.

[Ton03]   V. Toneev, E. Nikonov, B. Friman, W. Nörenberg, K. Redlich, *Strangeness production in nuclear matter and expansion dynamics.* Eur. Phys. J. C **32**, 399 (2003), doi:10.1140/epjc/s2003-01374-2, arXiv:hep-ph/0308088.

[VB98]   B. Vanderheyden, G. Baym, *Self-Consistent Approximations in Relativistic Plasmas: Quasiparticle Analysis of the Thermodynamic Properties.* J. Stat. Phys. **93**, 843 (1998), doi:10.1023/B:JOSS.0000033166.37520.ae.

[Ven92]   R. Venugopalan, M. Prakash, *Thermal properties of interacting hadrons.* Nucl. Phys. A **546**, 718 (1992), doi:10.1016/0375-9474(92)90005-5.

[vHe08]   H. van Hees, R. Rapp, *Dilepton radiation at the CERN super-proton synchrotron.* Nucl. Phys. A **806**, 339 (2008), doi:10.1016/j.nuclphysa.2008.03.009.

[VK72]   A. N. Vasil'ev, A. K. Kazanskii, *Legendre transforms of the generating functionals in quantum field theory.* Teor. Mat. Fiz. **12**, 352 (1972), doi:10.1007/BF01035606.

[VK73a] A. N. Vasil'ev, A. K. Kazanskii, *Equations of motion for a Legendre transform of arbitrary order.* Teor. Mat. Fiz. **14**, 289 (1973), doi:10.1007/BF01029302.

[VK73b] A. N. Vasil'ev, A. K. Kazanskii, *Convexity properties of Legendre transformations (variational methods in quantum field theory).* Teor. Mat. Fiz. **15**, 43 (1973), doi:10.1007/BF01028262.

[Wal02] W. Walter, *Analysis 2* (Springer, 2002), ISBN 9783540429531.

[Web99] F. Weber, *Pulsars as astrophysical laboratories for nuclear and particle physics* (CRC Press, 1999), ISBN 0750303328.

[Web05] F. Weber, *Strange quark matter and compact stars.* Prog. Part. Nucl. Phys. **54**, 193 (2005), doi:10.1016/j.ppnp.2004.07.001, arXiv:astro-ph/0407155.

[Wel82] H. A. Weldon, *Effective fermion masses of order gT in high-temperature gauge theories with exact chiral invariance.* Phys. Rev. D **26**, 2789 (1982), doi:10.1103/PhysRevD.26.2789.

[YHM95] K. Yagi, T. Hatsuda, Y. Miake, *Finite-Temperature Field Theory* (Cambridge University Press, 2005), ISBN 0521561086.

Acknowledgments

I would like to thank Prof. B. Kämpfer for accepting me as Ph.D. student and his encouraging assistance in elaborating this thesis. I also would like to thank my colleagues M. Bluhm, T. Hilger, H. Schade, D. Seipt, R. Thomas and the other members of our group and institute for extensive and fruitful discussions, helpful input and remarkable companionship.

The excellent working conditions which are due to the Institute of Theoretical Physics and the Research Center Dresden-Rossendorf are gratefully acknowledged. Especially the administrative support of A. Liebezeit, D. Schich and J. Steiner as well as G. Latus and G. Schädlich has enabled me to concentrate on the work at hand.

Without ranking names I would like to thank everyone in the heavy-ion physics community, where I felt welcome and which provided me with countless helpful discussions at conferences and workshops.

Most of all, I would like to thank my wife Constanze and my daughter Rahel for their love and understanding.

i want morebooks!

Buy your books fast and straightforward online - at one of world's fastest growing online book stores! Environmentally sound due to Print-on-Demand technologies.

Buy your books online at
www.get-morebooks.com

Kaufen Sie Ihre Bücher schnell und unkompliziert online – auf einer der am schnellsten wachsenden Buchhandelsplattformen weltweit! Dank Print-On-Demand umwelt- und ressourcenschonend produziert.

Bücher schneller online kaufen
www.morebooks.de

VDM Verlagsservicegesellschaft mbH
Heinrich-Böcking-Str. 6-8
D - 66121 Saarbrücken

Telefon: +49 681 3720 174
Telefax: +49 681 3720 1749

info@vdm-vsg.de
www.vdm-vsg.de

Printed by Books on Demand GmbH, Norderstedt / Germany